HarperCollinsPublishers
哈珀·柯林斯出版集团

双子星科普文库

THE CARTOON GUIDE TO ENVIRONMENT

爆笑科学漫画

环保超人

[美] 拉里·高尼克　爱丽丝·沃特瓦特◎著　叶云屏◎等译　刘夙◎校订
Larry Gonick & Alice Outwater

科学普及出版社
·北京·

图书在版编目（CIP）数据

环保超人：汉英对照 /（美）拉里·高尼克，（美）爱丽丝·沃特瓦特著；叶云屏等译.
—北京：科学普及出版社，2014.9（2020.4重印）
（爆笑科学漫画）
ISBN 978-7-110-08488-5

Ⅰ. ①环… Ⅱ. ①拉… ②爱… ③叶… Ⅲ. ①环境保护—青少年读物—汉、英 Ⅳ. ①X-49

中国版本图书馆CIP数据核字（2019）第074252号

策划编辑 郑洪炜
责任编辑 郑洪炜 李 洁
责任校对 王勤杰
责任印制 马宇晨
装帧设计 鑫联必升

出 版 科学普及出版社
发 行 中国科学技术出版社有限公司发行部
地 址 北京市海淀区中关村南大街16号
邮 编 100081
发行电话 010-62173865
投稿电话 010-63581070
网 址 http://www.cspbooks.com.cn

开 本 787mm×1092mm 1/16
字 数 300千字
印 张 17.75
印 量 19001—24000册
版 次 2014年9月第1版
印 次 2020年4月第6次印刷
印 刷 北京玥实印刷有限公司

书 号 ISBN 978-7-110-08488-5/X·52
定 价 49.80元

TO BOB, WITHOUT WHOM THIS BOOK WOULD NEVER HAVE BEEN WRITTEN

致鲍勃，没有他的鼓励，本书不可能完成。

•CONTENTS•

目 录

·CHAPTER 1·

FORESTS AND WATER

◆第一章◆

森林和水

OUR STORY BEGINS IN A PLACE THAT'S BEEN CALLED THE MOST FAR-FLUNG INHABITED ISLAND IN THE WORLD: ***EASTER ISLAND***, A 64-SQUARE-MILE SPECK IN THE PACIFIC OCEAN, 2300 MILES FROM ANYWHERE.

在世界上有人居住的最偏远的地方，有一个名叫复活节岛的小岛，我们的故事便从这个小岛讲起。复活节岛面积只有64平方英里（1英里＝1609.344米，下同），在浩瀚的太平洋上就像一个斑点，它距离最近的陆地有2300英里。

"HOW INAPPROPRIATE TO CALL THIS PLANET EARTH, WHEN CLEARLY IT IS OCEAN."

—ARTHUR C. CLARKE

"把这个星球叫作地球真不合适，这明明是一个水球！"

——阿瑟C·克拉克*

*阿瑟C·克拉克是英国著名科幻作家，未来学家。

REMOTE, BUT NOT ***DESERTED***... FROM TIME TO TIME VISITORS DROPPED BY... LIKE THE DUTCH ADMIRAL ***ROGGEVEEN*** IN 1722. ARRIVING ON ***EASTER SUNDAY***, HE NAMED THE ISLAND AFTER THE DATE OF ARRIVAL, AND LEFT THE FIRST WRITTEN ACCOUNT OF THE PLACE AND THE PEOPLE WHO LIVE THERE.

复活节岛虽然偏远，但并非**荒无人烟**……时常会有游客光顾……如1722年，荷兰海军上将**罗格文**就率领探险队来过这里。由于到达的那天正好是**复活节**，因此，罗格文便把这个岛命名为复活节岛，并留下了关于这个小岛及其居民的最早文字记录。

ACCORDING TO ROGGEVEEN AND OTHER 18TH-CENTURY REPORTS, SOME 3000 ISLANDERS EKED OUT A WRETCHED EXISTENCE BY FARMING BANANAS, SUGAR CANE, AND SWEET POTATOES FROM POOR, ROCKY SOIL. THE ONLY FRESH WATER CAME FROM MURKY LAKES INSIDE VOLCANIC CRATERS. THERE WAS SCARCELY A TREE ON THE ISLAND, AND THE PEOPLE WERE "SMALL, LEAN, TIMID, AND MISERABLE."

从罗格文的记录及其他18世纪的文献可以得知，当时岛上大约有3000名居民，他们在贫瘠多石的土地上靠种植香蕉、甘蔗和甘薯勉强维生。唯一的淡水取自火山口内混浊的湖水。岛上几乎看不到一棵树，当地居民“个子矮小、瘦骨嶙峋、胆小羞怯，一副可怜巴巴的样子。”

BUT AMIDST THE SQUALOR WERE SOME **SURPRISES...** ESPECIALLY SOME 800 ***MASSIVE STONE STATUES*** SCATTERED ACROSS THE ISLAND, SHOULDER TO SHOULDER, THEIR BACKS TO THE SEA. HOW WERE THEY CARVED? HOW QUARRIED? HOW MOVED? HOW ERECTED? AND BY WHOM?

但在这片肮脏贫瘠的土地上，却有着一些**神秘而奇异**的东西……最为瞩目的是岛上矗立的800多尊**巨大石雕**。它们肩并肩成行，背向大海而立。石像是如何雕凿出来的？如何采料？如何搬运？又如何竖立起来？这一切都是谁做的呢？

WHEN THE EUROPEANS ASKED WHERE THE STATUES HAD COME FROM, THE ISLANDERS GAVE THIS REPLY:

当欧洲来客问岛上居民石雕从何而来时，得到的回答是：

FAILING TO RECOGNIZE SARCASM WHEN THEY HEARD IT, THE EUROPEANS EMBARKED ON A SERIES OF WILD SPECULATIONS THAT HAVE CONTINUED TO THE PRESENT DAY: ***SPACEMEN*** SET UP THE STATUES WITH ***ANTI-GRAVITY DEVICES...*** A HIGHLY CIVILIZED ***LOST CONTINENT*** HAD SUNKEN INTO THE SEA, LEAVING ONLY EASTER ISLAND BEHIND... THEY WERE FLUNG INTO PLACE IN ONE PIECE BY ***VOLCANIC ERUPTIONS,*** ETC. ETC. ETC.

欧洲人并未听出话中的讽刺意味，因此开始了漫无边际的猜测，直到今天猜测依然在继续：是**外星人**利用**反地心引力装置**架起了这些石雕……这里曾是一块具有高度发达文明的**大陆**，但已沉入海底，唯有复活节岛残存下来……是**火山喷发**的威力将全部石雕抛到了这里……如此等等。

SINCE THE ISLANDERS HAD EITHER FORGOTTEN WHAT HAPPENED OR DIDN'T FEEL LIKE SHARING, IT WAS LEFT TO WESTERN SCIENTISTS AND HISTORIANS TO PIECE TOGETHER THE STORY WITH **CALIPERS, SHOVELS, MICROSCOPES,** AND **ETHNOGRAPHIC SURVEYS.**

AND HERE IS WHAT THEY FOUND OUT...
以下便是他们的发现……

岛上居民或许是已经忘了石雕从何而来，或许是不愿把这故事和人分享。既然如此，西方科学家和历史学家只好自己利用**卡尺**、**铁锹**、**显微镜**和**民族志调查**等科学手段来拼凑出这个故事。

AROUND THE YEAR 400, EASTER ISLAND WAS COLONIZED BY **POLYNESIANS.** VARIOUS FEATURES OF THE ISLANDERS' SKULLS, THEIR BLOOD TYPES, SOCIAL SYSTEM, LANGUAGE, AND CROPS ARE ALL POLYNESIAN.

大约公元400年，**波利尼西亚人**移居到复活节岛。岛上居民的颅骨、血型、社会制度、语言和农作物都具有波利尼西亚人的特征。

POLLEN SAMPLES TAKEN FROM LAKE BEDS SHOW THAT THE ISLAND WAS THEN THICKLY COVERED WITH VEGETATION. HACKING OUT CLEARINGS FROM THE JUNGLE, THE POLYNESIANS BUILT AND PLANTED, AND SOON THEY ENJOYED A TYPICAL POLYNESIAN DIET OF YAM, TARO, BREADFRUIT, BANANA, SUGAR, COCONUT, CHICKEN, AND POLYNESIAN RAT (SMALL AND TASTY!).

取自湖床的花粉试样显示，岛上曾经植被茂密。波利尼西亚人在丛林中开辟空地，建造房屋，种植作物。不久，他们便过上了典型的波利尼西亚人的生活，他们的食谱中包括山药、芋头、面包果、香蕉、蔗糖、椰子、鸡肉以及波利尼西亚鼠（个头小、味道美！）。

THEIR LIFE WAS RICH... THEIR BABIES THRIVED... THEY POPULATED THE ISLAND WITH LITTLE EFFORT... AND IN THEIR COPIOUS SPARE TIME, THEY CARVED STONE MONUMENTS, ESPECIALLY ***STATUES.***

这里的人们生活富足、人丁兴旺，在岛上悠闲自得地生活……他们把大量空闲时间用来雕刻纪念碑，尤其是**石雕**。

THE STATUES WERE MOVED AND SET UP WITH LOGS AND ROPE. RECENT EXPERIMENTS* HAVE PROVED THAT WHEN A STATUE IS STRAPPED UPRIGHT ONTO LOGS CARVED INTO SLED RUNNERS, AND RAISED ONTO A TRACK OF WOODEN ROLLERS, A COUPLE OF DOZEN PEOPLE COULD PULL IT EASILY!

*BY AMERICAN GEOLOGIST CHARLES LOVE.

*这些实验由美国地质学家查尔斯·洛夫完成。

这些石雕是用原木和绳索来搬运和放置的。最近的实验证明，如果将石雕竖立着绑在用原木雕成的雪橇状滑板上，地面上再用原木连起来铺成滑轨，只要几十个人便能轻松拉动这巨大的石雕！

AND BY THE YEAR 1400, THERE WAS SCARCELY A TREE LEFT STANDING ON EASTER ISLAND...

到了1400年，整个复活节岛上几乎再也看不到一棵树……

AND WHAT GOOD ARE TREES?? YOU MIGHT ASK... READ ON...

看到这里你也许会问，树木有什么好处呢？那就继续往下读吧……

THE WATER CYCLE 水循环

TO UNDERSTAND THE ROLE OF TREES, WE NEED TO LOOK AT THE WAY ***WATER*** MOVES AROUND THIS WATERY PLANET OF OURS.

要理解树木所起的作用，我们需要先看看**水**在我们这颗“水球”上的活动特点。

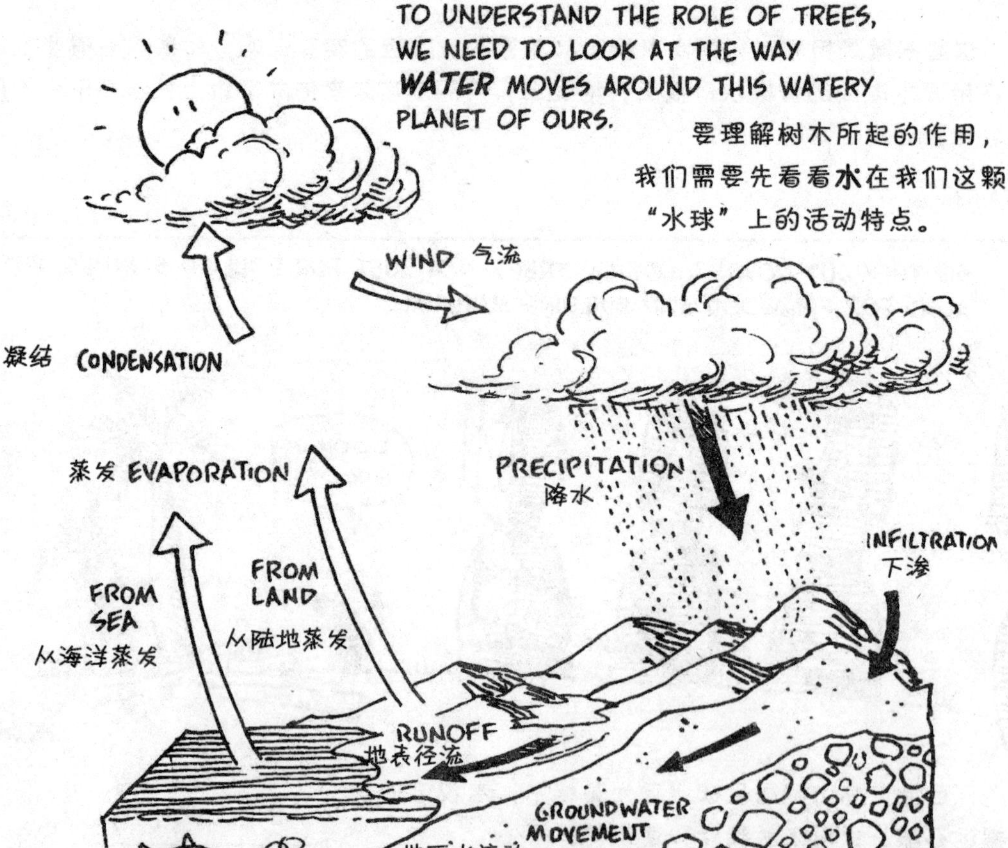

CLOUDS HOLD MOISTURE, WHICH FALLS AS RAIN (OR SNOW, BUT NOT IN POLYNESIA!). IT FALLS WITHIN A ***WATERSHED***, AN AREA THAT COLLECTS SMALL STREAMS INTO A MAJOR RIVER, AND ULTIMATELY RUNS TO THE SEA. WATER EVAPORATES FROM LAND AND SEA INTO THE AIR, WHERE IT CONDENSES INTO CLOUDS, AND THE CYCLE IS COMPLETED. WATER IN THE ATMOSPHERE IS FULLY REPLACED EVERY ***12 DAYS***.

云层中含有大量水汽，水汽凝结后降落下来形成雨或雪（不过在波利尼西亚是绝不会下雪的哦）。降雨的地方会形成一个**流域**，也就是很多小溪汇成大河的地方。河水最终流入大海。水分不断从陆地和海洋蒸发，进入大气，再次凝结成云，水循环便完成了。大气中的水每隔12**天**循环更新一次。

WHEN PRECIPITATION FALLS IN A FOREST, THE WATER HAS MANY OPTIONS: SOME BARELY PENETRATES THE GROUND BEFORE IT IS TAKEN UP BY ROOTS AND PASSED BACK TO THE AIR BY TRANSPIRATION (PLANT BREATHING). SOME GOES DEEPER, ALL THE WAY TO THE GROUNDWATER.

落到森林里的雨水去哪里了呢？这有很多种可能：有的还没等渗入地下，便被植物根系吸收，然后又通过蒸腾作用（植物也需要呼吸！）进入大气。有的渗到地下较深处，一直渗到地下水中。

(WHAT'S GROUNDWATER? DIG A DEEP ENOUGH HOLE IN THE GROUND, AND YOU'LL HIT WATER. THAT'S GROUNDWATER. THE TOP OF THE GROUNDWATER IS THE ***WATER TABLE***.)

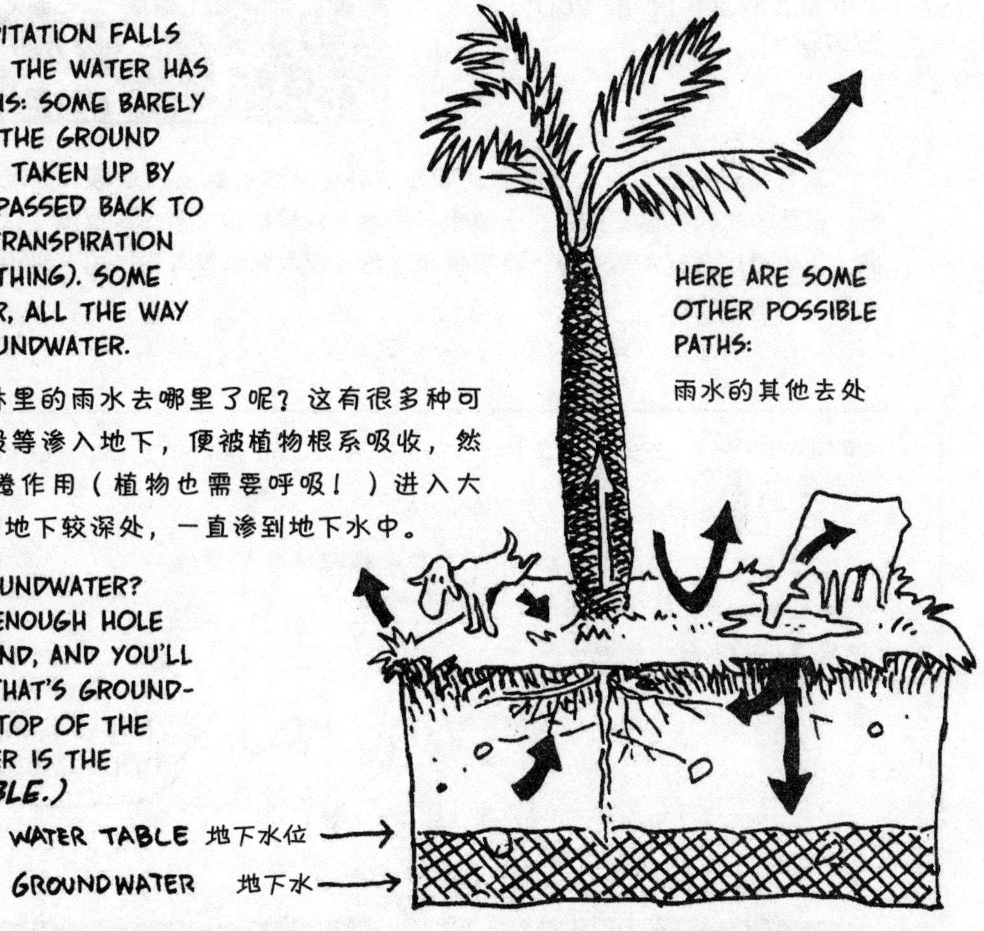

（什么是地下水呢？如果在地上挖一个深坑，就会有水渗出来，这就是地下水。地下水面也称**地下水位**。）

FOREST SOIL HOLDS A LOT OF WATER BECAUSE IT'S SO ***POROUS:*** A MIXTURE OF CLAY, SAND, AND DECAYING ORGANIC MATTER, THE SOIL IS HONEYCOMBED BY CHANNELS MADE BY ROOTS, BURROWING ANIMALS, AND FUNGI. THE TOP LAYERS TEEM WITH BACTERIA, WHICH BREAK DOWN ORGANIC COMPOUNDS INTO CHEMICAL NUTRIENTS THAT DISSOLVE IN WATER, DRIP DOWN, AND ARE TAKEN UP BY ROOT SYSTEMS.

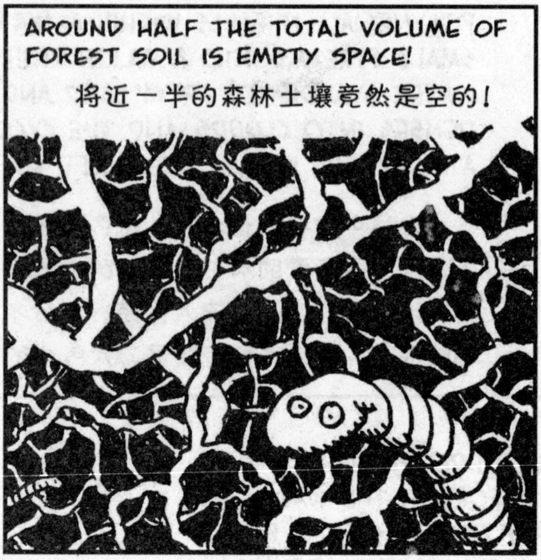

森林里的土壤含有大量水分，因为这种土壤含有黏土、沙砾和腐烂的有机物质，土质疏松，有很多**孔隙**。土壤里的树根、穴居动物、菌类等纵横交错，犹如蜂窝。土壤表层含有大量细菌，将有机化合物分解成化学营养素，溶入水中，再渗入深层，被根系吸收。

BY CONTRAST, OPEN LAND HAS LESS BIOLOGICAL ACTIVITY AND SO HOLDS LESS WATER.

相比之下，因为空旷的陆地上生物活动较少，所以这里储存的水分也就更少。

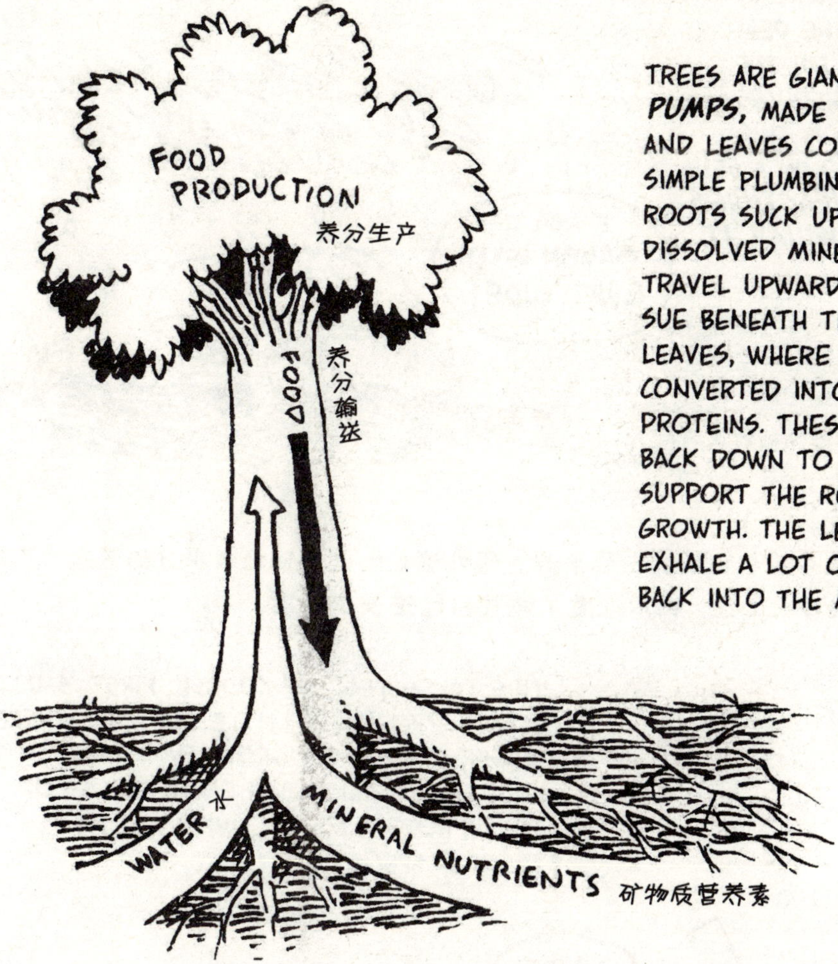

TREES ARE GIANT **WATER PUMPS**, MADE UP OF ROOTS AND LEAVES CONNECTED BY A SIMPLE PLUMBING SYSTEM. THE ROOTS SUCK UP WATER AND DISSOLVED MINERALS, WHICH TRAVEL UPWARD THROUGH TISSUE BENEATH THE BARK TO THE LEAVES, WHERE THEY ARE CONVERTED INTO SUGAR AND PROTEINS. THESE FOODS TRAVEL BACK DOWN TO THE ROOTS TO SUPPORT THE ROOTS' FURTHER GROWTH. THE LEAVES ALSO EXHALE A LOT OF WATER VAPOR BACK INTO THE ATMOSPHERE.

每棵树都像一台**巨型抽水机**，其组成部分树根和树叶由一个简单的管道系统连接。树根吸收水分和溶解了的矿物质，水分和矿物质由树皮下的组织向上输送给树叶，然后转化为糖分和蛋白质，成为树木的养分。这些养料又向下重新回到树根，为树根提供继续生长的养分。树叶也会呼出大量水汽，散发到大气中。

UNDERGROUND, HUNDREDS OF MILES OF TREE ROOTS ANCHOR THE TREE TO THE GROUND AND HOLD THE SOIL IN PLACE. MARVELOUS THING, A TREE!

在地下，总长达数百英里之多的树根使大树牢牢固定在土壤中。仅仅是一颗树，却如此神奇！

FALLING LEAVES MAKE UP A GOOD PART OF THE DEAD ORGANIC MATTER ON THE GROUND. SHADE REDUCES EVAPORATION FROM THIS MATERIAL, AND SO, IN A SENS THE TREE ***MANUFACTURES THE SOIL IN WHICH IT GROWS.***

落叶是地表土中有机物质的主要来源。落叶遮盖地面，使地表水分的蒸发作用减弱。因此，从某种意义上说，树木**营造了适宜自己生长的土壤**。

THIS TAKES TIME, OF COURSE. FIRST, SMALL PLANTS PRODUCE A LITTLE TOPSOIL, WHERE BIGGER PLANTS CAN GROW. EVENTUALLY, ENOUGH NUTRIENT-RICH EARTH BUILDS UP TO SUPPORT A FOREST, WITH ALL ITS TEEMING SPECIES.

当然，这一过程需要很长时间。最先是矮小的植物在地上生长会产生薄薄的一层土，支持稍大些的植物生长。依此类推，最终积聚形成厚厚的土壤，其营养之丰富足以支持整片森林以及森林里所有物种的生长。

WHAT HAPPENED ON EASTER ISLAND? WHEN A TREE IS CUT DOWN AND ITS ROOTS DIE, THE TOPSOIL LOSES ITS ANCHOR. 4 TO 5 FEET OF EARTH MAY EVENTUALLY BE WASHED AWAY, AND THERE IS NO QUICK WAY TO REPLACE IT.

那复活节岛上何以落得如此贫瘠？

当树木被砍伐，树根死亡，表土层就失去了固定作用，四五英尺*厚的土壤被雨水冲走，流失的土壤也没有快速有效的方法得到补充。

*1英尺＝0.3048米，下同。

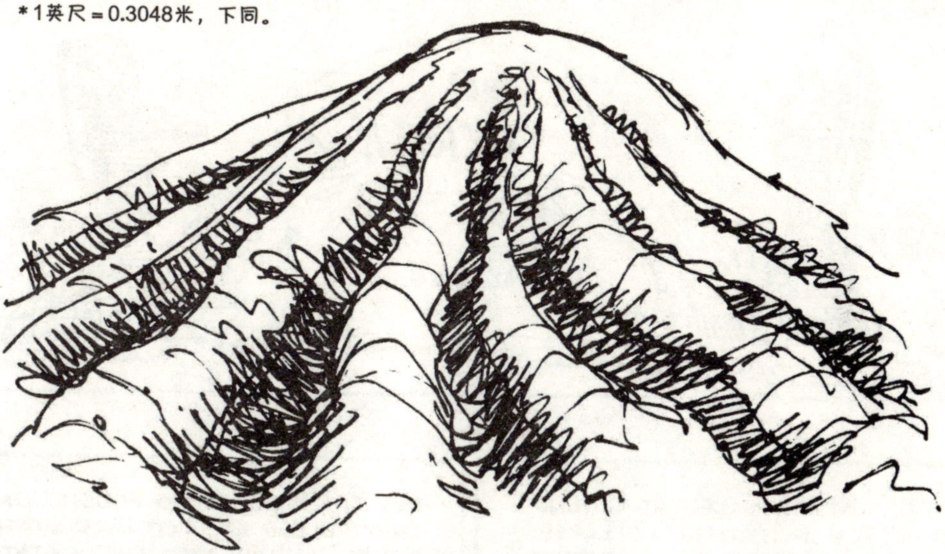

WITHOUT FORESTS TO ABSORB RAIN AND REPLENISH GROUNDWATER, THE ISLAND'S STREAMS AND SPRINGS DRIED UP... THE AIR BECAME LESS HUMID, AND RAINFALL DIMINISHED. AS FERTILE TOPSOIL ERODED, CROP YIELDS FELL... THERE WAS NO WOOD FOR HOUSES... NO FIBERS FOR FISHING NETS OR SAILCLOTH, NO LOGS FOR CANOES...

失去了树木和森林，雨水不能被土壤吸收，地下水得不到补充，岛上的溪流和泉水枯竭了……空气不那么湿润了，降雨量也减少了。肥沃的表土层受到侵蚀，农作物减产了……盖房子的木材没有了……做渔网或帆布的纤维也没有了，做小船的木头也没有了……

RIVALRY FOR RESOURCES LED TO PERMANENT WARFARE. UNFORTUNATELY, CLAN PRESTIGE WAS DISPLAYED BY ***ERECTING STATUES***, SO THE LAST TREES WERE PROBABLY CUT DOWN IN A FRENZIED EFFORT TO SHOW OFF...

资源争夺还引发了没完没了的战争。更糟糕的是，显示部落威风的办法是**凿建雕像**，因此，岛上最后几棵树很可能是各个部落为了炫耀威风而在极度疯狂中被砍掉的……

THE POPULATION PEAKED AT AROUND 7000 PEOPLE IN 1550 AND COLLAPSED SO QUICKLY THAT SOME 400 STATUES WERE LEFT UNFINISHED IN THE QUARRIES.

岛上人口在1550年达到高峰——将近7000人。之后迅速减少，由于缺乏人工，石场上还有四百多尊雕像没完工。

BUT THE RIVAL CLANS FOUGHT ON... THEY PULLED EACH OTHERS' STATUES DOWN... AND BY 1860, EVERY STATUE ON THE ISLAND HAD BEEN TOPPLED.

但部落间的战争还在继续……他们相继摧毁对方的石雕……到1860年，岛上所有的石像都被推翻在地。

THE POINT OF THIS STORY IS NOT THAT THE PEOPLE OF EASTER ISLAND WERE SOMEHOW STRANGE, SILLY, OR DIFFERENT FROM ANYONE ELSE. QUITE THE CONTRARY: LIKE THE REST OF US, THEY WERE ***CREATURES OF HABIT,*** AND THEIR WAY OF LIFE—FARMING, FORESTRY, BUILDING, AND DISPLAY—WAS HARD TO CHANGE.

讲这个故事的目的并不是为了说明复活节岛上的居民是多么怪异、愚蠢或者多么异常。恰恰相反，他们和我们大家一样，都有自己的**生活习惯**。他们的生活方式，如种地、伐木、建筑和炫耀等，都很难改变。

EASTER ISLAND IS VERY SMALL. FROM ITS SUMMIT YOU CAN SEE THE WHOLE THING. THE PERSON WHO CUT DOWN THE LAST TREE MUST HAVE KNOWN THERE WERE NO MORE—AND STILL HE CUT IT DOWN.

复活节岛很小。站在岛的最高处，整个岛屿的面貌一目了然。砍掉最后一棵树的那个人一定也知道，岛上再也没有树木了，但他还是挥起了斧头。

OUR PLANET, WHILE MUCH LARGER, IS STILL FINITE. LIKE THE ISLANDERS, WE TOO CAN SEE IT ALL, AND LIKE THE ISLANDERS WE HAVE NO MEANS OF ESCAPE. IS IT STILL POSSIBLE THAT WE CAN TAKE STOCK OF OUR RESOURCES AND CHANGE OUR HABITS IN TIME TO AVOID THE FATE OF EASTER ISLAND?

尽管我们的地球比复活节岛要大得多，但资源仍然是有限的。像岛上的居民一样，我们也能看到地球上的一切，我们也无处可逃。如果我们从现在开始珍惜资源，改变生活习惯，从而避免复活节岛居民的悲惨命运重蹈覆辙，还来得及吗？

CHAPTER 2

MORE CYCLES

第二章

更多循环

THE WATER CYCLE, OR **HYDROLOGIC CYCLE** AS SCIENTISTS CALL IT, IS ONLY ONE OF THE CHEMICAL CYCLES ENGINEERED BY LIFE. THE EARTH'S ELEMENTS CONSTANTLY MOVE AROUND IN ENORMOUS, INTERLOCKING LOOPS, CREATED AND DIRECTED BY LIVING ORGANISMS.

水循环，科学家也称之为**水文循环**，它只是生物系统中众多的化学循环之一。地球上的元素在巨大的、错综复杂的链环中循环移动，这些链环是由生命有机体创造并控制的。

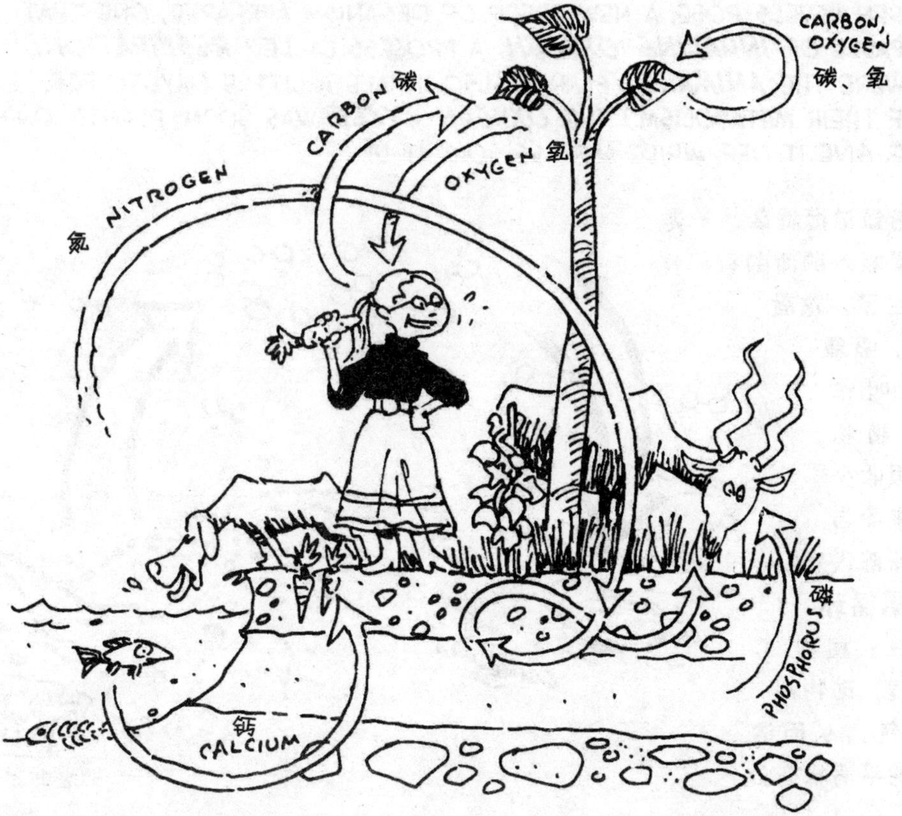

TAKE ***OXYGEN,*** FOR EXAMPLE... WHEN THE EARTH WAS YOUNG, SOME 4 BILLION YEARS AGO, THERE WAS LITTLE FREE OXYGEN IN THE ATMOSPHERE. THEN LIFE APPEARED, AND SOON THERE WERE MICROBES CALLED ***CYANOBACTERIA,*** THEN LARGER CREATURES CALLED ***PLANTS,*** THAT EXHALED PURE MOLECULAR OXYGEN (O_2) AS A WASTE BY-PRODUCT.

以**氧元素**为例……40亿年前，当地球刚刚形成时，大气中很少有自由氧元素。随后出现了生命：一类名叫**蓝菌**的微生物很快出现，之后出现了较大些的生物——**植物**，释放出纯净的氧气（O_2），可对植物来说，这是一种废气。

AS OXYGEN LEVELS ROSE, A NEW BREED OF ORGANISM APPEARED, ONE THAT WAS CAPABLE OF ***INHALING OXYGEN,*** A PROCESS CALLED ***RESPIRATION.*** THESE WERE THE ***ANIMALS.*** (PLANTS ALSO INHALE A LITTLE OXYGEN FOR PART OF THEIR METABOLISM.) THE ***OXYGEN CYCLE*** WAS BORN: PLANTS, ON BALANCE, GIVE IT OFF, WHILE ANIMALS SOAK IT UP.

随着氧浓度提高，一类能够将氧吸入的新的有机体产生了，这就是**动物**。吸氧的过程叫作**呼吸**（植物也可以吸收少量的氧来参与自身的新陈代谢）。**氧循环**由此产生：植物释放氧气，动物则吸收氧气，从而达到氧气的平衡状态。

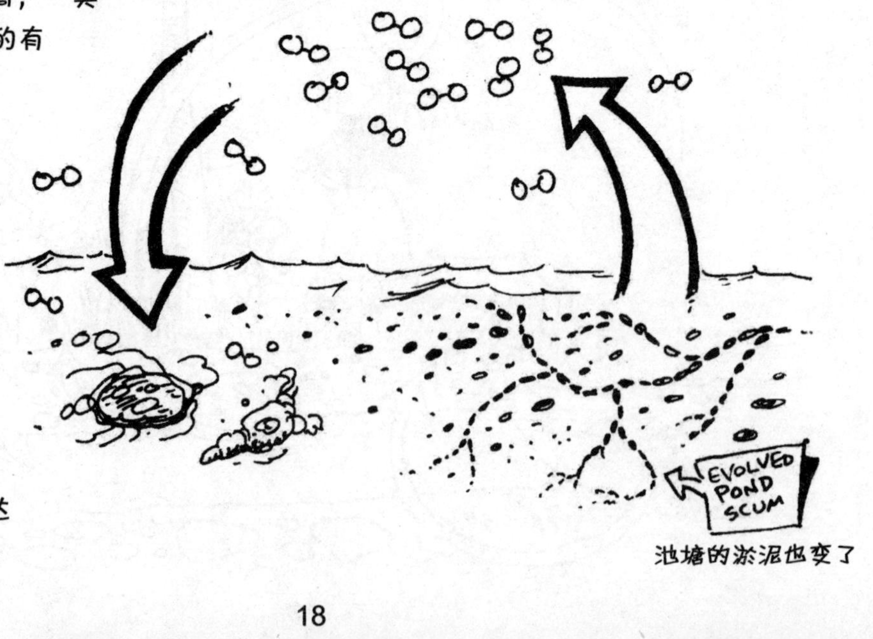

ANIMAL POPULATIONS SWELLED TO CONSUME THE ABUNDANT OXYGEN, UNTIL A LIMIT WAS REACHED, AND THE WHOLE SYSTEM CAME INTO ***DYNAMIC BALANCE.*** FOR THE PAST 2 BILLION YEARS, PLANTS AND ANIMALS HAVE HELD THE OXYGEN LEVEL AT A VERY STEADY ***21%*** OF THE TOTAL ATMOSPHERE.

动物数量急剧增加，消耗了大量氧，最终达到一定极限，整个系统处于**动态平衡**状态。在过去的20亿年里，植物和动物使大气中的氧含量始终保持在21%的水平。

IMPORTANT NOTE: AS IT HAPPENS, ANIMALS EXHALE ***CARBON DIOXIDE GAS*** (CO_2), WHICH PLANTS ABSORB AS A RAW MATERIAL FOR BUILDING THEIR CARBON-BASED TISSUES. PLANTS ARE CALLED

PRODUCERS

BECAUSE THEY MAKE ("PRODUCE") ORGANIC MATTER DIRECTLY FROM CO_2 AND OTHER CHEMICALS.

值得注意的是：动物呼出的**二氧化碳**（CO_2）恰好被植物吸收，作为原材料形成自身的碳基组织。植物被称为**生产者**，因为它们能够直接利用CO_2和其他化学物质制造有机物。

ANIMALS, ON THE OTHER HAND, ARE

CONSUMERS:

THEY GET FOOD BY EATING PLANTS OR OTHER ANIMALS!

而动物是**消费者**，以植物或其他动物为食。

ONE INTERPRETATION OF THE OXYGEN STORY IS THAT THE ***TOTALITY OF LIFE ON EARTH REGULATED THE ATMOSPHERE:*** LIFE CREATED AND MAINTAINED CONDITIONS FAVORABLE TO THE GROWTH OF MORE LIFE. COULD THIS BE TRUE OF OTHER CYCLING CHEMICALS AS WELL?

关于对氧元素的解释有一种观点认为：**地球上的万物共同调节了大气**。生物创造并维持了利于更多生命成长的条件。其他化学物质循环也是这样吗？

IN THE 1970S, THE BRITISH CHEMIST JAMES LOVELOCK AND AMERICAN BIOLOGIST LYNN MARGULIS PROPOSED THE ***GAIA*** (GUY-UH) ***HYPOTHESIS:*** THEY SUGGEST THAT THE WORLD IS AN INTERCONNECTED, BIOLOGICAL BEING, WHOM THE SCIENTISTS CALLED GAIA AFTER THE GREEK EARTH GODDESS.

20世纪70年代，英国化学家詹姆斯·拉夫洛克和美国生物学家林恩·马古利斯提出了**盖亚假说**，即地球是一个相互联系的生物体，科学家以希腊神话中的大地女神盖亚命名，将这一生物体称为盖亚。

THE GAIA HYPOTHESIS SAYS THAT LIVING SYSTEMS ACT AS ***ENVIRONMENTAL REGULATORS,*** ENSURING THAT CHEMICAL LEVELS STAY WITHIN LIMITS CONGENIAL TO GAIA'S HEALTH.

盖亚假说认为，生命系统作为**环境的调控者**，确保化学物质含量维持在能够保证盖亚健康的限度内。

FOR ANOTHER EXAMPLE, ***SALT*** IS CONTINUOUSLY WASHED FROM LAND TO SEA, YET THE LEVEL OF OCEAN SALINITY HAS BARELY CHANGED SINCE LIFE BEGAN. LOVELOCK SUGGESTS THAT CORAL REEFS EXTRACT SALT FROM THE OCEAN BY WALLING OFF ***EVAPORATION POOLS,*** WHERE SALT IS ISOLATED FROM THE SEA.

再以盐为例吧，**盐**不断从陆地冲入海洋，然而海水盐度自生命起源起就没有多大变化。拉夫洛克认为，是珊瑚礁把海水分隔，形成一个个**蒸发池**，提取了海水中的盐分。

IN LOVELOCK AND MARGULIS' VIEW, THE ABIOTIC (NONLIVING) WORLD—THE ***ATMOSPHERE*** (GAS), ***HYDROSPHERE*** (WATER), AND ***GEOSPHERE*** (THE HARD STUFF)—IS REGULATED BY THE ***BIOSPHERE*** (THE LIVING WORLD). THE ***ECOSPHERE*** IS THE INTERACTION BETWEEN THE BIOSPHERE AND ALL THE WORLD'S ELEMENTS.

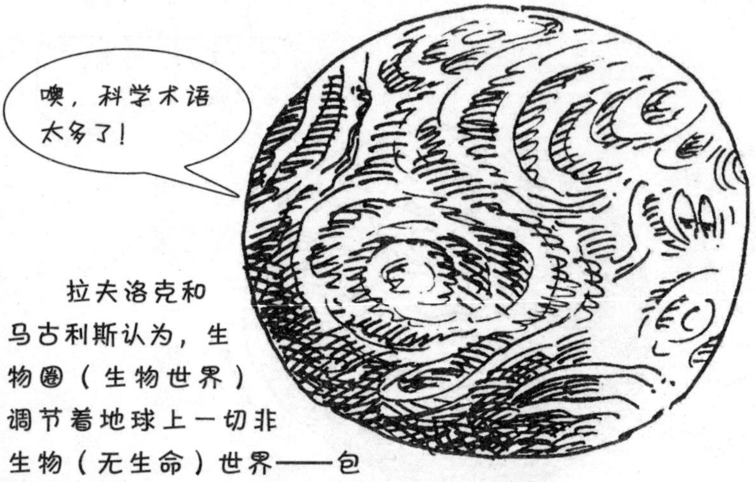

拉夫洛克和马古利斯认为，生物圈（生物世界）调节着地球上一切非生物（无生命）世界——包括**大气圈**（大气）、**水圈**（水）、**岩石圈**（岩石）。生物圈和地球上其他元素相互作用形成生态圈。

THE GAIA THEORY SAYS THAT WHENEVER AN ABIOTIC CONDITION BEGINS TO GET OUT OF HAND, LIFE RESPONDS IN SOME WAY THAT PULLS THE CONDITION BACK UNDER CONTROL, DAMPING DOWN THE FLUCTUATION.

盖亚假说提出，每当一种非生物条件失控时，生物总会以某种方式作出反应，将它控制，平息波动。

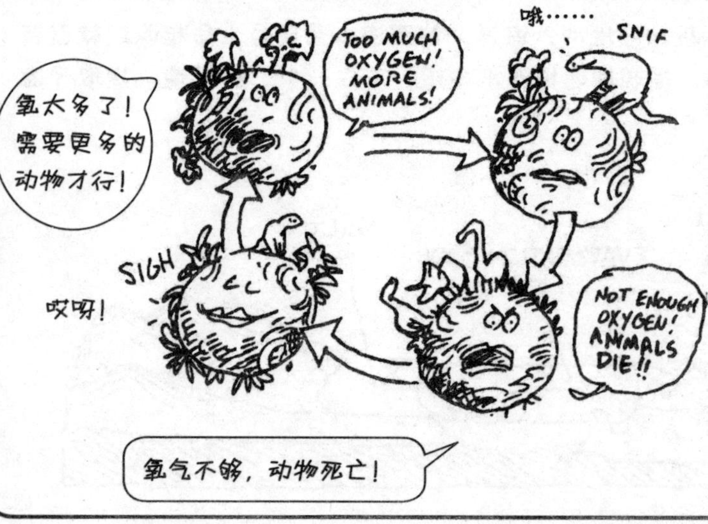

IN THIS WAY, LIVING SYSTEMS CREATE ***FEEDBACK LOOPS.*** THEY RECEIVE INFORMATION ABOUT THE ECOSPHERE AND REACT TO IT. THIS REACTION PRODUCES A CHANGE, AND ANOTHER REACTION...

通过这种方式，生命系统建立起**反馈环路**，收集生态圈信息并作出反应。反应引起变化，又带来新的反应……

LOVELOCK AND MARGULIS BELIEVE (OR HOPE) THAT THESE FEEDBACK LOOPS ARE MAINLY STABILIZING AND REGULATORY. A CONDITION MAINTAINED IN THIS WAY IS CALLED ***HOMEOSTASIS***, A ***DYNAMIC STEADY STATE***, FULL OF FLUX AND CHANGE, BUT ULTIMATELY TUNED TO OPTIMIZE CONDITIONS FOR LIFE.

拉夫洛克和马古利斯相信（或希望），这些反馈环路主要起着稳定和调节的作用。在稳定和调节作用下维持的状态叫作稳态，即**动态稳定状态**，虽然不断波动和变化，但最终能够得到调节，使生物的生存条件得到优化。

THIS VERY ATTRACTIVE IDEA CONTAINS A GOOD DEAL OF TRUTH... BUT AS WE'LL SEE LATER, FEEDBACK LOOPS CAN ALSO BE ***DESTABILIZING***, AND THE GAIA HYPOTHESIS REMAINS CONTROVERSIAL.

这个理论很吸引人，也的确很有道理……但我们会发现，反馈环路也有可能起到**破坏稳定**的作用。因此，盖亚假说仍有争议。

BUT TRUE OR FALSE, THE GAIA HYPOTHESIS REMINDS US TO THINK OF THE BIOSPHERE AS A SINGLE GIGANTIC SYSTEM.

不过，无论盖亚假说是对还是错，它至少提醒我们要把生物圈看成一个单一的庞大系统。

OF THE 90+ NATURALLY OCCURRING ELEMENTS ON EARTH, ABOUT 40 ARE REQUIRED BY LIVING ORGANISMS. SINCE VIRTUALLY NO NUTRIENTS COME FROM OUTER SPACE, THESE ATOMS ARE CONSTANTLY RECYCLED. A CARBON ATOM ON YOUR NOSE MAY HAVE ONCE BEEN A TRICERATOPS' TOENAIL.

地球上存在着九十多种自然元素，其中约有40种是生物体所必需的。既然没有养分来自外太空，这些原子就处于不断的循环中。你鼻子上的一个碳原子，或许曾经还是三角龙的一个脚趾甲呢！

THE ***MACRONUTRIENTS***—THOSE NUTRIENTS WE NEED IN LARGE QUANTITIES—ARE CARBON, OXYGEN, HYDROGEN, NITROGEN, PHOSPHORUS, SULFUR, CALCIUM, AND POTASSIUM.. THESE ELEMENTS MAKE UP MORE THAN 95% OF THE MASS OF ALL ORGANISMS, AND THEY MOVE AROUND A LOT. FOR EXAMPLE:

大量元素是指那些我们大量需要的营养元素，如碳、氧、氢、氮、磷、硫、钙、钾……这些元素构成了95%以上的生物有机体，并且还不停运动。例如：

1.碳原子进入体内，随血液流动成为脚趾甲的一部分。

4.碳原子经过更多变化，落到了你的鼻子上。

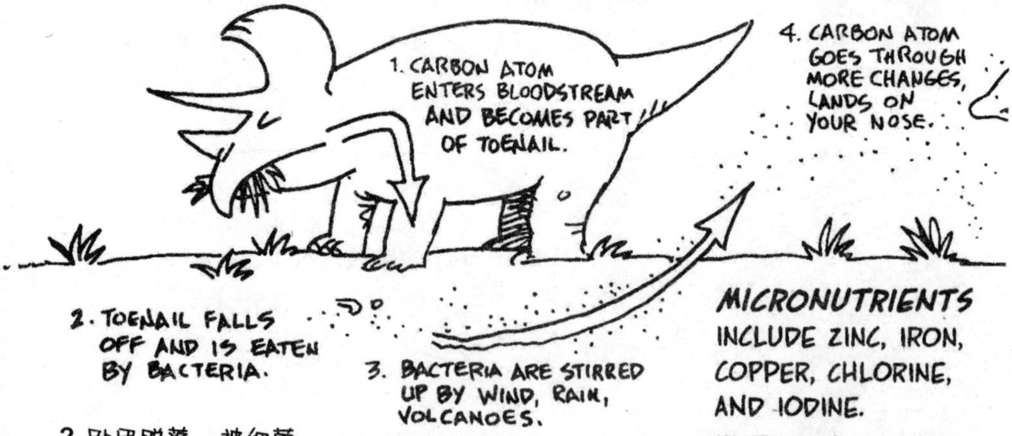

MICRONUTRIENTS INCLUDE ZINC, IRON, COPPER, CHLORINE, AND IODINE.

微量元素包括锌、铁、铜、氯和碘。

2.趾甲脱落，被细菌分解。

3.细菌随风、雨或火山喷发而搅动。

WHEN A CREATURE DIES AND IS BURIED IN SEDIMENTS, ITS ELEMENTS SINK BELOW GROUND AND ARE LOST FOR EONS—UNLESS SCAVENGERS EXTRACT KEY ELEMENTS FROM THE CORPSE AND SPREAD THEM AROUND.

当生物死亡并掩埋在沉积物里时，它身上的元素会沉入地下亿万年，除非食腐生物把尸体里的重要元素提取和传播。

IF ORGANIC MATTER DOES SINK UNDERGROUND, IT USUALLY RECYCLES VERY SLOWLY. IT MAY NOT COME UP AGAIN FOR MILLIONS OF YEARS, UNTIL EROSION OR GEOLOGICAL UPHEAVAL EXPOSES IT.

有机物若沉入地下，循环通常会非常缓慢。也许数百万年后，由于侵蚀作用或地质变化才会重现地面。

GASEOUS ELEMENTS—OXYGEN, NITROGEN, AND CARBON (IN CARBON DIOXIDE)--CYCLE RAPIDLY, BECAUSE THE AIR IS FULL OF THEM AND BLOWS THEM AROUND. THEY MAY RECYCLE INTO LIFE AND OUT AGAIN WITHIN HOURS OR DAYS.

气态元素——氧、氮和二氧化碳中的碳等遍布在空气中并随空气四处流动，因此循环很快。仅需几小时或几天，它们就可能在生物体内外进出循环一次。

BETWEEN THESE EXTREMES ARE MEDIUM-TERM CYCLES OF ELEMENTS CIRCULATING THROUGH THE HYDROSPHERE AND OTHER LIVING ORGANISMS.

介于这两个极端之间的是那些通过水圈或其他有机体来进行中期循环的元素。

IN SHORT, THE RATE OF BIOGEOCHEMICAL RECYCLING DEPENDS LARGELY ON THE ACCESSIBILITY OF THE ELEMENT'S MAJOR RESERVOIR.

总之，生物地质化学的循环速度在很大程度上取决于元素是否可以很快接近各自的主要储存库。

EACH ELEMENT HAS A CYCLE, BUT WE DO NOT FULLY UNDERSTAND HOW ALL OF THEM ARE REGULATED.

每种元素都有各自的循环系统，但我们并不完全了解它们都是如何被调控的。

BEST KNOWN ARE THE THREE BIG ONES:

我们最了解的元素有三大类：

NITROGEN

(AN ESSENTIAL INGREDIENT OF PROTEINS AND DNA): ALMOST 80% OF THE ATMOSPHERE IS PURE NITROGEN (N_2), A FORM WHICH IS USELESS TO MOST ORGANISMS.

氮是蛋白质和DNA的重要组成成分。空气中将近80%的成分是高纯度氮气（N_2），但氮气的这种形态对多数有机体来说并没有用处。

FIXED NITROGEN IS TAKEN UP AND USED BY OTHER PLANTS, WHICH MAY IN TURN BE EATEN BY ANIMALS.

固定的氮被其他植物吸收利用，而这些植物又可能被动物吃掉。

BUT CERTAIN BACTERIA, ESPECIALLY THOSE LIVING ON ROOT NODULES OF PEAS, BEANS, ALFALFA, ALDER TREES, ETC., CAN FIX NITROGEN, I.E., CONVERT IT TO NITRATES (NO_3) AND AMMONIA (NH_3).

但有些细菌，尤其是那些生长在豌豆、蚕豆、苜蓿和桤木树等根瘤处的菌类，能够固氮，也即将其转化为硝酸盐（NO_3）或氨（NH_3）。

CARBON,

碳，在空气中，碳以二氧化碳（CO_2）的形式存在；在水里，是溶解了的二氧化碳；在陆地上，则以碳酸钠、石灰石和白垩等碳酸盐岩石和其他化合物的形式存在。

CARBON IS FOUND IN THE AIR, AS CARBON DIOXIDE, CO_2, IN WATER, AS DISSOLVED CO_2, AND IN THE GROUND, AS CARBONATE ROCKS (SODA, LIMESTONE, CHALK) AND OTHER COMPOUNDS.

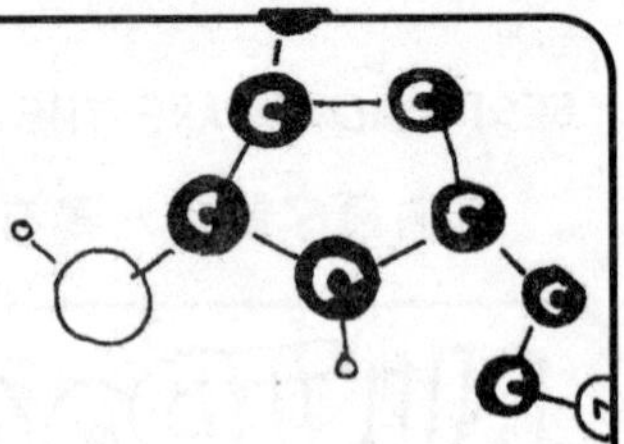

THE BASIC BUILDING BLOCK OF ALL LIFE:

碳是构成生命的基本元素。

GREEN PLANTS USE PHOTOSYNTHESIS TO BUILD SUGARS AND CARBOHYDRATES FROM CO_2 AND WATER, WITH SUNLIGHT PROVIDING THE ENERGY.

在光合作用下，绿色植物利用阳光提供的能量，将二氧化碳和水转化成糖和碳水化合物。

THESE COMPOUNDS ARE EATEN BY CONSUMERS, SO THE CARBON MOVES FROM PLANT TO ANIMAL.

这些化合物被消费者吃掉，碳也就随之从植物转移到动物身上。

SOME RETURNS TO THE ATMOSPHERE AS EXHALED CO_2.

火山喷发时，会将大量二氧化碳释放到大气中。

DISSOLVED CO_2 AND OTHER CARBON COMPOUNDS

溶解的二氧化碳和其他含碳化合物

BOTH PLANTS AND ANIMALS DIE AND DECOMPOSE, RELEASING MOST CO_2 BACK INTO THE AIR, BUT ALSO DRIPPING OTHER CARBONACEOUS COMPOUNDS INTO THE SOIL.

动植物死后分解，将大量二氧化碳释放回空气，同时也使另一些含碳化合物进入土壤。

UNDERGROUND, SOME CARBON IS SEQUESTERED IN ***FOSSIL FUELS:*** PEAT, COAL, OIL, AND NATURAL GAS.

在地下，一些碳形成了泥炭、煤、石油和天然气等**化石燃料**。

PHOSPHORUS

IS A KEY ELEMENT IN METABOLIC ENERGY TRANSFERS, BONES AND TEETH, AND OTHER TISSUES AND MOLECULES. IT MOVES IN A ***SEDIMENTARY*** CYCLE.

磷是新陈代谢中能量迁移的一种关键元素，也是骨骼、牙齿、其他组织和分子中的关键元素，以**沉积型**循环移动。

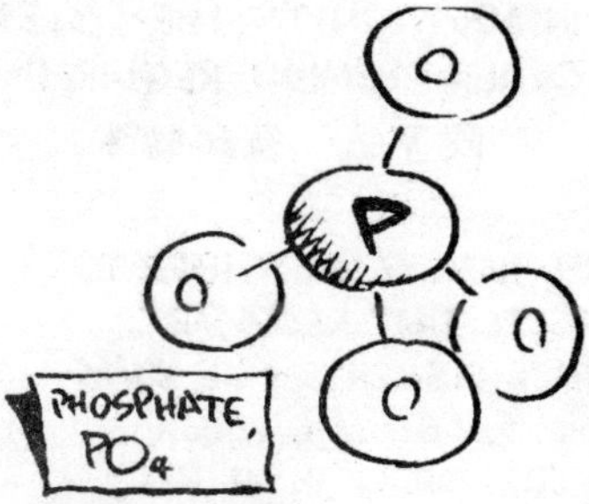

IN THE WATER, FISH GET PHOSPHORUS BY EATING PLANKTON AND THEN REDEPOSIT IT ON THE OCEAN FLOOR WHEN THEY DIE. FISH AND BIRDS ALSO MOVE IT AROUND.

在水中，鱼类通过捕食浮游生物获取磷，死亡后使磷沉积在海底。磷也会随鱼类和鸟类四处迁移。

THE MAIN RESERVOIR OF PHOSPHORUS IS ***PHOSPHATE ROCKS*** IN SEDIMENTARY BEDS LAID DOWN IN THE PAST. WATER DISSOLVES THE PHOSPHORUS OUT OF THE ROCKS INTO THE SOIL, WHERE PLANTS ABSORB IT THROUGH THEIR ROOTS. ANIMALS EAT THE PLANTS AND SHED PHOSPHORUS BY EXCRETION OR DECOMPOSITION.

磷的主要储存库是**磷矿石**，位于过去沉淀下来的沉积岩中。水溶解矿石中的磷，再流入土壤，被植物根系吸收。动物吃掉植物后通过排泄或分解，再次将磷分离出来。

AS PARTICLES SETTLE, PHOSPHORUS SINKS INTO THE DEEP OCEAN, WHERE IT CAN REMAIN FOR AGES.

随着微粒沉淀，磷也沉降到海底深处，并在那里被保存很长时间。

IN ADDITION TO THE CYCLES OF CARBON, NITROGEN, AND PHOSPHORUS, EVERY OTHER ELEMENT REQUIRED BY LIFE HAS ITS OWN CYCLE.

除了碳、氮和磷外，生物需要的每一种其他元素都有各自的循环系统。

LIFE HAS TO WORK HARD TO ENSURE THAT NECESSARY TRACE ELEMENTS LIKE ***ZINC*** AND ***SELENIUM,*** WHICH OCCUR RARELY IN NATURE, REMAIN AVAILABLE AND AREN'T BURIED IN SEDIMENTS OR THE OCEAN FLOOR.

有些必要元素，如**锌**和**硒**这些微量元素，在自然界中含量很少，因此，生命体必须不停地工作，以确保它们的循环，不让它们淹没在沉积岩中或是海底。

ALL THESE CYCLES INTERACT WITH EACH OTHER IN COMPLICATED WAYS WE HAVE BARELY BEGUN TO UNDERSTAND!!

所有这些循环系统都以复杂的方式相互作用，对此我们却知之甚少！

LIFE, TAKEN AS WHOLE WORLDWIDE SYSTEM, SIMULTANEOUSLY CREATES, MAINTAINS, AND DEPENDS ON A RICH SET OF INTERLOCKING CYCLES. AND A GOOD THING, TOO! OUR LIVES LITERALLY DEPEND ON ADEQUATE SUPPLIES OF CHEMICAL INGREDIENTS, FROM AMINO ACIDS TO ZINC, AND THESE ARE SPREAD AROUND THE WORLD BY THE ACTION OF COUNTLESS LIVING THINGS.

作为一个全球性系统，生命体同时创造、维持并依赖于一系列错综复杂的循环。这是一件好事！我们的生命依赖足够的化学成分，无论是氨基酸还是锌都是如此，这些元素通过无数生物的活动四处播散。

CAN THERE BE A BETTER REASON FOR PRESERVING THE WEB OF LIFE?

还有更好的理由来保护这张生物网吗？

CHAPTER 3

EVOLVING SYSTEMS, STRUGGLING INDIVIDUALS

第三章

进化的系统，挣扎的个体

UNDER THE INFLUENCE OF LIFE, THE EARTH'S CHEMICAL CYCLES WERE CREATED, MODIFIED, AND MAINTAINED OVER THE EONS. COMPLEX WEBS OF ORGANISMS MOVE PRECIOUS TRACE ELEMENTS AROUND. SOME HAVE EVEN ARGUED THAT THE EARTH BEHAVES LIKE A GIANT LIVING THING, REGULATING CHEMICAL CYCLES TO ITS OWN ADVANTAGE.

AT LEAST, THAT'S HOW IT LOOKS FROM A DISTANCE, WHEN BIO-, GEO-, AND HYDROCHEMICAL CYCLES ARE VIEWED AS ***SYSTEMS.*** FROM THE STANDPOINT OF THE ***INDIVIDUAL ORGANISM,*** HOWEVER, THINGS LOOK A LITTLE DIFFERENT. BORN INTO THIS VAST FLUX OF CYCLING CHEMICAL RESOURCES, THE INDIVIDUAL HAS A SIMPLER, MORE SELFISH GOAL:

在生命体的影响下，地球的化学循环得以建立、调节和维持几十亿年。错综复杂的生物网中珍贵的微量元素在不断运动。有观点认为，地球就像一个巨大的生物体，调节着化学循环向利于自己的方向发展。

从宏观上看，生物循环、地质循环和水文循环都可视为**系统**。然而，从**生物个体**的角度来看，情况略有不同。它们生来就处于这个庞大的化学资源循环运动中，目标更加简单——只要利于自己生存即可。

THE INDIVIDUAL ORGANISM STRIVES TO *EAT* AND *AVOID BEING EATEN* LONG ENOUGH TO *REPRODUCE SUCCESSFULLY* (I.E., SO THE OFFSPRING SURVIVE).

生物个体拼命让自己**吃饱**，同时**避免**自己**被**其他生物**吃掉**，以求获得足够长的时间来进行**成功的繁殖**，让后代存活。

ALTHOUGH THE INDIVIDUAL IS PART OF THE SYSTEM, THE INDIVIDUAL'S NEEDS ARE NOT NECESSARILY THE SAME AS THE SYSTEM'S NEEDS. FOR EXAMPLE, AN EASTER ISLANDER NEEDED ONLY *ONE TREE AT A TIME* FOR FUEL OR BUILDING MATERIAL, WHILE THE SOCIAL SYSTEM NEEDED A *FOREST* TO RUN SUCCESSFULLY.

虽然个体是系统的一部分，但两者的需求并非完全一致。例如，一位复活节岛居民**每次**只需要**一棵树**来做燃料或建筑材料，而扩大到社会系统，却需要整片**树林**才能成功运行。

INDIVIDUALS ARE BORN INTO A WEB OF SYSTEMS—FAMILY, COMMUNITY, SPECIES, BIOSPHERE—THAT CREATE OPPORTUNITIES AND IMPOSE LIMITS. AT THE SAME TIME, THESE SYSTEMS ARE CREATED BY THE ACTION OF ALL THEIR MYRIAD INDIVIDUALS.

个体生来就处于网络系统中——家族、群落、物种、生物圈，系统既给个体创造了生存的机会，却也为个体施加了束缚。与此同时，系统也是由所有形形色色个体的运动所创建的。

IN THE INTERPLAY OF INDIVIDUAL AND SYSTEM, **EVOLUTION** TAKES PLACE.

在个体和系统的相互作用中，**进化**产生了。

INDIVIDUALS COMPETE WITH EACH OTHER FOR RESOURCES. EVERY CREATURE TRIES TO SUSTAIN ITS OWN LIFE LONG ENOUGH TO REPRODUCE, AND IF THAT MEANS SOMEBODY ELSE GOES HUNGRY, TOO BAD!

个体之间相互夺取资源。每种生物都试图长期生存，繁衍后代。如果一种生物的生存意味着另一种生物要挨饿，那就太糟糕了!

TIGER SALAMANDER LARVA EATING ITS COMPETITOR

幼小的虎螈正在吃它的竞争对手

THIS DOESN'T MEAN THAT CREATURES DON'T COOPERATE ALSO! SOMETIMES COOPERATING WITH OTHERS IS THE MOST EFFECTIVE WAY TO SURVIVE.

这并非是说生物之间就没有合作。有时候，合作反倒是最好的生存方式。

NOT ALL INDIVIDUALS ARE ALIKE: EACH ONE HAS A SLIGHTLY DIFFERENT COMBINATION OF GENETIC TRAITS, AND SOME OF THESE COMBINATIONS GIVE THEIR OWNERS A ***SELECTIVE ADVANTAGE***: THE LUCKY ORGANISM IS BETTER ABLE TO GET FOOD, EVADE PREDATORS, WITHSTAND HEAT AND COLD, REPRODUCE, ETC.

并不是所有的个体都具有相同的特点，每种个体的遗传特征组合都会略有不同，其中有些组合给其主人**选择优势**：这些幸运的个体更有能力获取食物、逃避捕食者、忍耐严寒酷暑、繁衍后代等。

THE RESULT IS ***DIFFERENTIAL REPRODUCTION***: THE FAVORED INDIVIDUALS BREED MORE OFFSPRING... SOME OF THEIR OFFSPRING INHERIT THE GOOD GENES, SO THEY BREED MORE PRODUCTIVELY TOO. AFTER SEVERAL GENERATIONS, THE BETTER-ADAPTED TYPE MAKES UP MOST OF THE POPULATION.

结果就是**差异生殖**。具有选择优势的个体会繁育出更多的后代……后代中有一些成员继承优良基因，也大量繁殖。几代之后，适应性较强的种类占了种群的多数。

SOMETIMES THE DESCENDANTS WILL EVOLVE INTO AN ENTIRELY NEW SPECIES.

有时候，它们的后代会完全进化成为一个新的物种。

WHAT IS A SPECIES? GENERALLY SPEAKING, A SPECIES CONSISTS OF ORGANISMS THAT CAN MATE WITH EACH OTHER... BUT THE DEFINITION IS SOMEWHAT FUZZY, BECAUSE SOME ORGANISMS CAN MATE ACROSS SPECIES, AND WHEN YOU START TALKING ABOUT ORGANISMS LIKE ***BACTERIA***, OTHER CRITERIA MUST BE FOUND.

物种是什么呢？一般来说，物种是指能够互相交配的生物体……但这种定义有点模糊，因为有些生物可以跨物种交配。当考虑到**细菌**这类生物体时，必须采用其他标准。

EXACTLY HOW SPECIATION OCCURS IS NOT WELL UNDERSTOOD. MOST BIOLOGISTS BELIEVE IN

allopatric

("OTHER PLACE") SPECIATION: A SMALL POPULATION BECOMES GEOGRAPHICALLY ISOLATED IN SOME WAY. BREEDING ONLY AMONG THEMSELVES, ITS MEMBERS EVOLVE AWAY FROM THE ANCESTRAL TYPE. HUMANS AND CHIMPANZEES DIVERGED, IT IS BELIEVED, BECAUSE THE ANCESTRAL SPECIES WAS DIVIDED BY AFRICA'S GREAT RIFT VALLEY.

人们并不十分清楚物种究竟是如何形成的。多数生物学家认可**异域**物种形成的说法，认为小的种群由于地理隔离而只能在内部繁殖，在演化过程中种群成员偏离了其祖先的类型。他们认为，人类和黑猩猩得以分化，是由于非州大裂谷将他们的祖先隔离开来了。

Sympatric

("SAME PLACE") SPECIATION CAN INVOLVE ***SEASONAL*** OR ***HABITAT ISOLATION***—POTENTIAL MATES AREN'T IN THE SAME PLACE AT THE SAME TIME—OR ***BEHAVIORAL ISOLATION***, FOR EXAMPLE, WHEN A COURTSHIP RITUAL DEVELOPS THAT APPEALS TO SOME BUT NOT TO ALL.

同域物种的形成是因为**季节性**或**生境隔离**，即那些潜在的配偶并不出现在同一时间和同一地点；或者是因为**行为隔离**，比如一种求偶方式只能吸引部分异性而非全部。

IN ONE WAY OR ANOTHER, THE WORLD IS DIVIDED INTO MILLIONS OF SPECIES.

无论如何，整个世界上有数以百万计物种。

WITHIN A SPECIES, A **POPULATION** IS A GROUP OF INDIVIDUALS THAT ARE ACTUALLY AVAILABLE TO EACH OTHER FOR MATING. A SPECIES MAY BE SPREAD ACROSS THE GLOBE, WHILE A POPULATION TENDS TO CLUSTER TOGETHER.*

在同一物种内，**种群**是指能够彼此进行交配的个体的集合。一个物种可能会遍布全球生活，但一个种群往往聚居在一起*。

*EXCEPTIONS, AS ALWAYS, EXIST!

*当然，凡事都有个例外！

THE STUDY OF HOW POPULATIONS RISE AND FALL IS CALLED

POPULATION DYNAMICS,

AN INEXACT SCIENCE WITH LOTS OF GRAPHS.

研究种群盛衰状况的科学称为**种群动态学**，是一种运用大量图表的不太精确的研究。

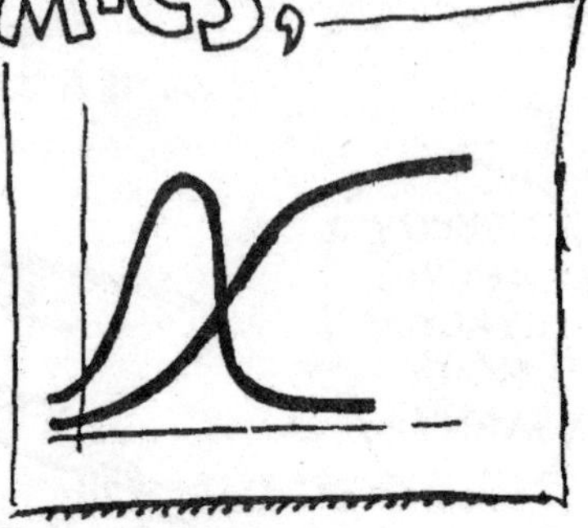

FOR NOW, LET'S JUST SKIM OVER A FEW KEY CONCEPTS.

现在，我们一起来看几个关键词吧！

BIOTIC POTENTIAL

IS THE MAXIMUM GROWTH RATE OF A POPULATION, OFTEN WRITTEN r_{max}. IT DEPENDS ON THE NUMBER OF OFFSPRING, THEIR AVERAGE SURVIVAL RATE, AND HOW EARLY AND OFTEN REPRODUCTION TAKES PLACE. r_{max} IS HARD TO MEASURE OUTSIDE THE LAB.

生物潜能是指一个种群的最大生长率，通常用r_{max}表示，通过对后代的数量、平均成活率、最早生育时间和生育频率统计得到。因此,它很难在实验室外测得。

REPRODUCTIVE STRATEGIES

ARE SUPPOSED TO ENSURE THAT BIRTHS EXCEED DEATHS, SO THAT THE POPULATION WILL INCREASE UNLESS CHECKED IN SOME WAY.

繁殖策略可以确保出生人数超过死亡人数，这样人口才会增长，除非用某种方法使之得到抑制。

ENVIRONMENTAL RESISTANCE

INCLUDES FACTORS THAT LIMIT POPULATION. THESE *LIMITING FACTORS* DETERMINE THE *CARRYING CAPACITY*, OR THE NUMBER OF INDIVIDUALS IN A POPULATION THAT CAN BE SUSTAINED INDEFINITELY IN A GIVEN ECOSYSTEM. THIS IS ANOTHER HARD NUMBER TO SPECIFY.

环境阻力指限制种群增长的因素。这些限制因子决定了**承载率**，即某一特定生态系统能够保证种群中可以长期持续生存的个体数。很难说这个数量具体是多少。

LIMITING FACTORS COME IN TWO FLAVORS: EXTRINSIC AND INTRINSIC.

限制因子包括外在的和内在的两种。

EXTRINSIC LIMITING FACTORS COME FROM OUTSIDE THE POPULATION. THEY CAN BE ***ABIOTIC*** (NONLIVING), SUCH AS AVAILABILITY OF CHEMICAL NUTRIENTS, THE LEVEL OF LIGHT, WATER, ETC. TOO MUCH OR TOO LITTLE OF ANY ONE ABIOTIC FACTOR CAN LIMIT POPULATION, EVEN IF ALL OTHER FACTORS ARE AT OPTIMAL LEVELS.

外在限制因子来自种群外部，可以是化学营养元素的可利用性、光照程度、水分等**非生物**因子。即便其他所有因子都维持在适宜水平，任何一种非生物因子过多或过少都会限制种群生长。

BIOTIC EXTRINSIC FACTORS INCLUDE FOOD SUPPLY, PREDATORS, DISEASE, AND THE GENERAL SUITABILITY OF THE ORGANIC ENVIRONMENT.

生物的外在限制因子包括食物供给、捕食者、疾病和有机环境的整体适宜性。

INTRINSIC LIMITING FACTORS COME FROM WITHIN THE POPULATION: ***REPRODUCTIVE RATE, ADAPTABILITY,*** AND EVEN ***TERRITORIALITY***: GROUPS OR INDIVIDUALS MAY STAKE OUT A TERRITORY, FROM WHICH OTHERS IN THE SPECIES ARE EXCLUDED. THIS DETERMINES A POPULATION DENSITY AND HENCE A MAXIMUM LEVEL.

内在限制因子来自种群内部，如**繁殖率**和**适应能力**，甚至**领域性**——即有些群体或个体会标出一块专属领地，杜绝物种内其他群体或个体进入。这就决定了种群密度和种群极限水平。

SOCIAL HIERARCHY CAN ALSO LIMIT POPULATION: IN MANY SPECIES, A DOMINANT MALE EJECTS LOWER-STATUS MALES FROM THE GROUP AND MAY EVEN KILL THEIR OFFSPRING.

社会等级之分也会给种群带来限制：在许多物种内，占优势地位的雄性会将其他地位较低的雄性从群体中驱逐出去，甚至杀掉它们的后代。

DIFFERENT SPECIES ADOPT DIFFERENT REPRODUCTIVE STRATEGIES, DEPENDING ON WHAT LIMITING FACTORS APPLY. THE TWO MOST EXTREME STRATEGIES ARE:

根据所受到的限制因子，物种不同，采用的繁殖策略也不同。最极端的两种繁殖策略是：

r SELECTION: HAVE MASSES OF OFFSPRING IN ONE REPRODUCTIVE SHOT, AND THEN DIE. THIS STRATEGY WORKS WELL IN AN UNCROWDED ENVIRONMENT, WHERE POPULATION MAY EXPAND RAPIDLY, A HAZARDOUS ENVIRONMENT WHERE FEW SURVIVE, OR A RAPIDLY CHANGING ENVIRONMENT WHERE SWIFT ADAPTATION IS NEEDED.

r选择：这种繁殖策略在以下情况中表现良好：不拥挤的环境，这时种群可快速扩大；少有个体能有活的危险环境；需要快速适应的骤变的环境。

K SELECTION: AN INDIVIDUAL REPRODUCES REPEATEDLY, HAS FEW OFFSPRING EACH TIME, CARES FOR THE YOUNG, AND HAS HIGH EXPECTATION FOR THEIR SURVIVAL. THIS STRATEGY OFTEN WORKS WELL WHEN POPULATION IS NEAR THE ENVIRONMENT'S CARRYING CAPACITY, BUT IT CAN BE RISKY. IF POPULATION DROPS TOO FAR, K SELECTION IS A SLOW WAY TO BUILD IT BACK UP.

K选择：个体反复繁殖，但每次繁殖的数量很少。这样，通过关心照料幼体，后代的存活率就高。这种策略通常是在种群接近环境承载力时行之有效，但仍有一定风险。如果种群数量锐减，恢复过程则非常缓慢。

* * * * * * * * * * *

ORANGUTANS, WITH JUST ONE OFFSPRING EVERY FIVE YEARS, ARE A PRIME EXAMPLE.

* * * * * * * * * * * *

那些5年才育有一个后代的红毛猩猩，就是最典型的例子。

THEN THERE IS A RANGE OF STRATEGIES IN BETWEEN.

还有一系列繁殖策略介于两这者之间!

GIVEN THESE INFLUENCES ON POPULATION, WE CAN DRAW GRAPHS SHOWING PATTERNS OF POPULATION GROWTH.

考虑到这些因素对种群的影响，我们可以用示意图来展示种群的成长模式。

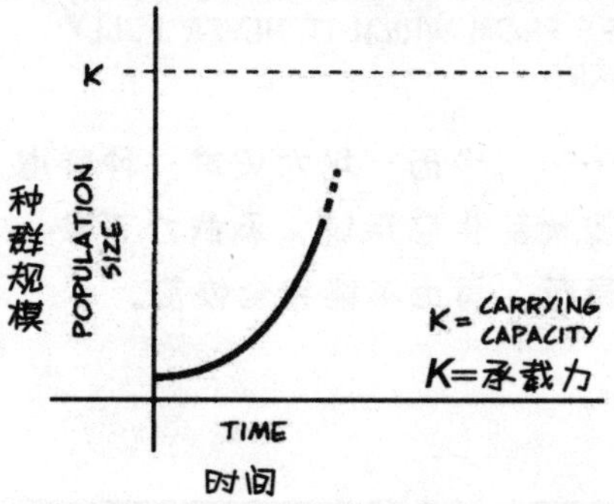

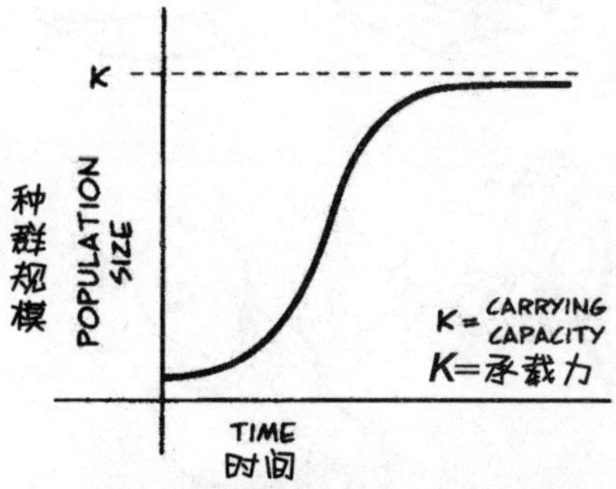

THERE MAY BE ENOUGH INTRINSIC AND EXTRINSIC CHECKS TO CAUSE THE POPULATION TO LEVEL OFF AT K.

可能有太多内在和外在制约导致种群规模偏离承载水平。

WHEN A POPULATION IS SMALL COMPARED TO THE CARRYING CAPACITY, IT GROWS MORE OR LESS RAPIDLY, DEPENDING ON ITS INTRINSIC RATE OF REPRODUCTION. WHAT HAPPENS NEXT DEPENDS ON MANY FACTORS.

当种群规模小于承载力时，种群增长速度会根据其内在繁殖率适当加快。接下来的情况取决于众多因素。

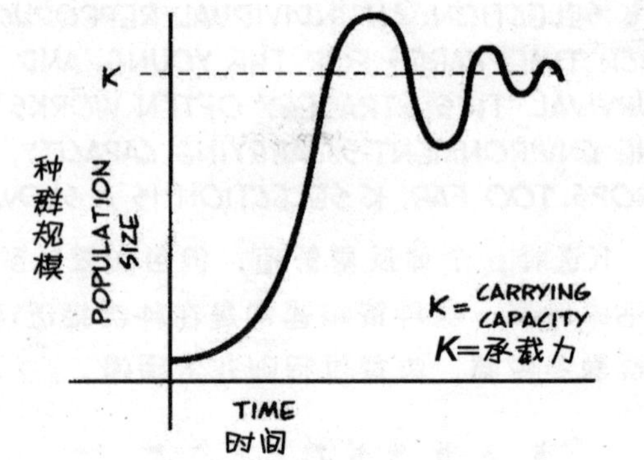

MORE TYPICALLY, POPULATION OVERSHOOTS K, IN WHICH CASE IT MAY WOBBLE BACK TO AN EQUILIBRIUM LEVEL...

更为典型的是，如果种群规模超出了承载力K，种群数量就会不断变动，重新达到平衡状态……

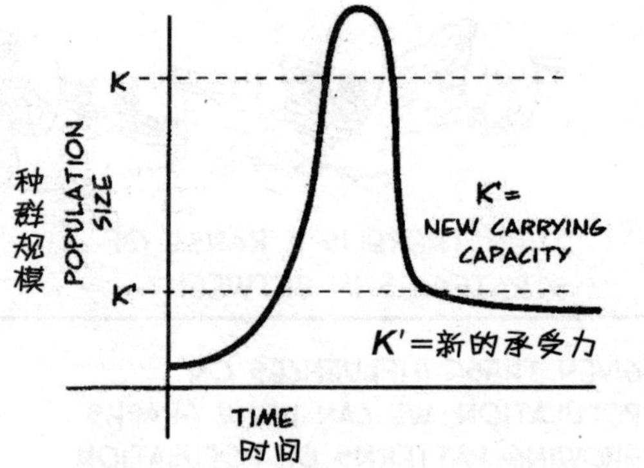

...OR HAVE A CATASTROPHE: POPULATION GROWTH DISRUPTS THE ECOSYSTEM SO BADLY THAT ITS CARRYING CAPACITY IS REDUCED, AND THE SPECIES SUFFERS A DIE-OFF FROM WHICH IT NEVER FULLY RECOVERS.

……或经历一场大灾难：种群增长严重扰乱生态系统，承载力下降，物种衰落，再也不能完全恢复。

ALTHOUGH EVOLUTION IS DRIVEN BY COMPETITION FOR RESOURCES, THE LAST GRAPH ON THE PREVIOUS PAGE SHOWS THE DANGER FOR A SPECIES OR POPULATION THAT COMPETES TOO "SUCCESSFULLY" IN THE GAME OF LIFE.

虽然进化是在资源争夺的推动下发生的，但上页最后那张图显示，那些在竞争中太“成功”的物种或种群，也会面临危机。

IN THE WEB OF LIFE, EACH SPECIES DEPENDS ON THE SURVIVAL OF MANY OTHERS.

在整个生物网中，每个物种都要依赖于大量其他物种才能生存。

A STRIKING EXAMPLE IS GIVEN BY ***COEVOLUTION:*** TWO (OR MORE) SPECIES MAY ***COEVOLVE*** TO TAKE ADVANTAGE OF EACH OTHER. FLOWERS, FOR INSTANCE, HAVE STRUCTURES THAT RUB OFF POLLEN ON NECTAR-SEEKING BEES, WHILE BEE HIND LEGS ARE PERFECTLY SUITED TO ACCUMULATE POLLEN. AS THE BEE FLITS FROM FLOWER TO FLOWER, IT FERTILIZES THE PLANTS, WHICH DEPEND COMPLETELY ON BEES FOR REPRODUCTIVE SUCCESS.

最明显的例子就是**协同进化**，即两个或更多物种之间相互利用**共同进化**。例如，植物分泌花蜜吸引蜜蜂前来采蜜，同时将花粉沾到蜜蜂身上，而蜜蜂的后足则非常适合收集花粉。蜜蜂从一朵花飞到另一朵花，使植物受精。这些植物完全依靠蜜蜂进行繁殖。

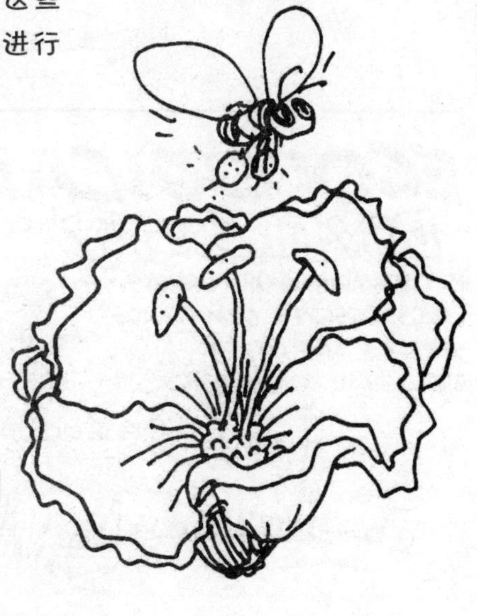

BIODIVERSITY COMES IN THREE FLAVORS:

生物多样性有三个方面：

GENETIC DIVERSITY

IS DIVERSITY WITHIN A SPECIES. DIFFERENT INDIVIDUALS DO HAVE SMALL DIFFERENCES IN THEIR GENETIC MAKEUP. GENETIC DIVERSITY IS HEALTHY BECAUSE IT MAKES A SPECIES MORE ADAPTABLE TO ENVIRONMENTAL CHANGE.

遗传多样性指同一物种内的多样性。不同个体在遗传构造上确实存在少许差异。遗传多样性使得物种可以更加适应环境变化，因此是有益的。

SPECIES DIVERSITY

IS THE VARIETY OF SPECIES IN A COMMUNITY (INCLUDING THE BIGGEST COMMUNITY, THE WHOLE EARTH).

物种多样性是指在一个群落里物种的丰富多样的程度（整个地球作为最大的群落也包括在内）。

ECOLOGICAL DIVERSITY

IS THE VARIETY OF FORESTS, LAKES, DESERTS, GRASSLANDS, STREAMS, AND OTHER BIOLOGICAL COMMUNITIES.

生态多样性是指各种各样的森林、湖泊、荒漠、草地、溪流以及其他生态群落的多样程度。

EVOLUTION NOT ONLY GENERATES GENETIC DIVERSITY BUT ALSO DEPENDS ON IT. INDIVIDUAL DIFFERENCES WITHIN A POPULATION INCREASE THE LIKELIHOOD THAT SOME VARIANTS WILL SURVIVE CHANGES IN THE ENVIRONMENT. SPECIES AND ECOLOGICAL DIVERSITY ALSO GIVE LIFE RESILIENCY.

进化不仅能够产生遗传多样性，也依赖遗传多样性。种群内的个体差异使得其在环境变化中存活下去的可能性增大。物种多样性和生态多样性也加强了生命的韧性。

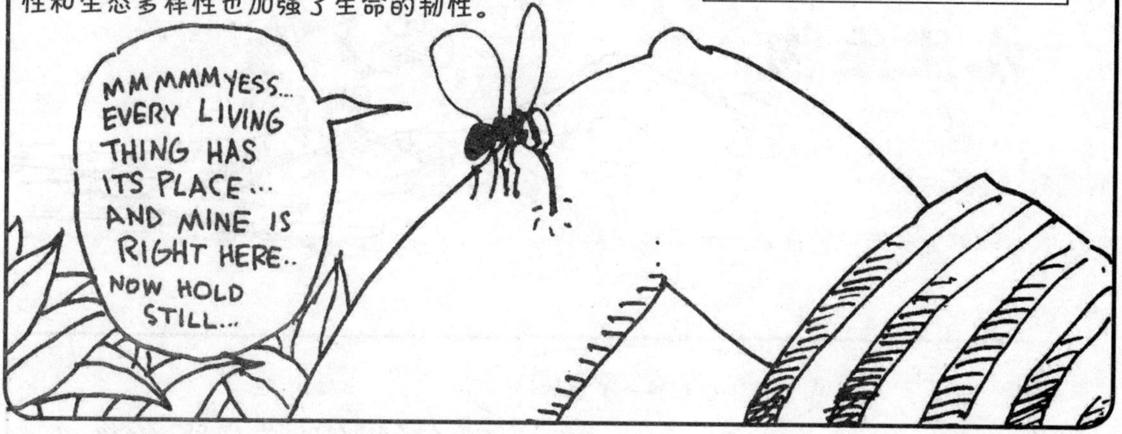

SINCE THESE LOCAL ECOLOGIES REGULATE THE CIRCULATION OF RESOURCES, LET'S SPEND SOME TIME REVIEWING THE WORLD'S LIVING COMMUNITIES, AS THEY SHRINK UNDER THE STEADY PRESSURE OF OUR OWN SPECIES' K-STRATEGIC EXPANSION.

既然这些局部生态调节着资源循环，那我们就来回顾一下地球上的生物群落吧！这些生物群数量因我们人口的K策略式增长而日益萎缩。

CHAPTER 4

COMMUNITIES WET...

◆第四章◆

水生生物

LET'S START IN THE OCEAN, WHERE LIFE BEGAN.

我们的故事从生命的摇篮——海洋讲起。

THE OCEAN COVERS OVER ***70%*** OF THE EARTH AND IS CONSTANTLY FLOWING. OCEAN CURRENTS MOVE MASSES OF WATER LONG DISTANCES: THE WARM GULF STREAM ALONE CARRIES ***50 TIMES MORE WATER*** THAN ALL THE WORLD'S RIVERS COMBINED. SINCE WATER HOLDS HEAT, OCEAN CURRENTS DISTRIBUTE SOLAR ENERGY, WHILE ITS SHEER BULK STABILIZES CLIMATIC CHANGE. THE OCEAN ALSO SERVES AS A RESERVOIR OF DISSOLVED GASES THAT REGULATES THE COMPOSITION OF THE ATMOSPHERE.

地球表面70%以上是海洋。海水不断流动，洋流使大规模海水向遥远的海区流动，例如墨西哥湾暖流使流入遥远海区的水量达到陆地上所有河流总量的**五十多倍**。由于水能储存热量，因此，洋流将太阳能传输并分配到不同地方，这些能量的大部分都用来使气候变化保持稳定。海洋中还储存了大量溶解气体，调节着大气构成。

OF THE 1.4 MILLION SPECIES THAT HAVE BEEN CLASSIFIED TO DATE, ONLY 250,000, OR 35%, LIVE IN THE OCEAN. THIS IS NOT BECAUSE THE OCEANS ARE LESS DIVERSE, BUT BECAUSE WE HAVE SPENT LESS TIME THERE LOOKING FOR NEW SPECIES. DESPITE OUR IGNORANCE, THERE ARE A FEW THINGS WE CAN SAY ABOUT UNDERWATER COMMUNITIES.

在迄今分类的140万个物种中，只有35%（约25万）的物种生活在海洋里。不是因为海洋中的生物缺乏多样性，而是因为我们并没有花太多时间去发现新的物种。虽然我们掌握的海洋生物的知识有限，但还是略知一二。

FOR THE MOST PART, AQUATIC LIFE PRACTICES r SELECTION: ADULTS BREED PROFUSELY, AND THE YOUNG RARELY REACH MATURITY. THIS IS THE HABIT OF MOST FISH AND EVERYTHING LOWER DOWN THE EVOLUTIONARY SCALE. MAMMALS LIKE SEALS, DOLPHINS, AND WHALES ARE THE EXCEPTIONAL K-BREEDERS.

在多数情况下，水生生物实行r选择：成体一次繁殖出大量幼体，而这些幼体多数都会夭折。这是多数鱼类和所有比鱼类的进化水平还低的低级生物的习性。海豹、海豚、鲸等哺乳动物例外，它们采用个体反复繁殖策略。

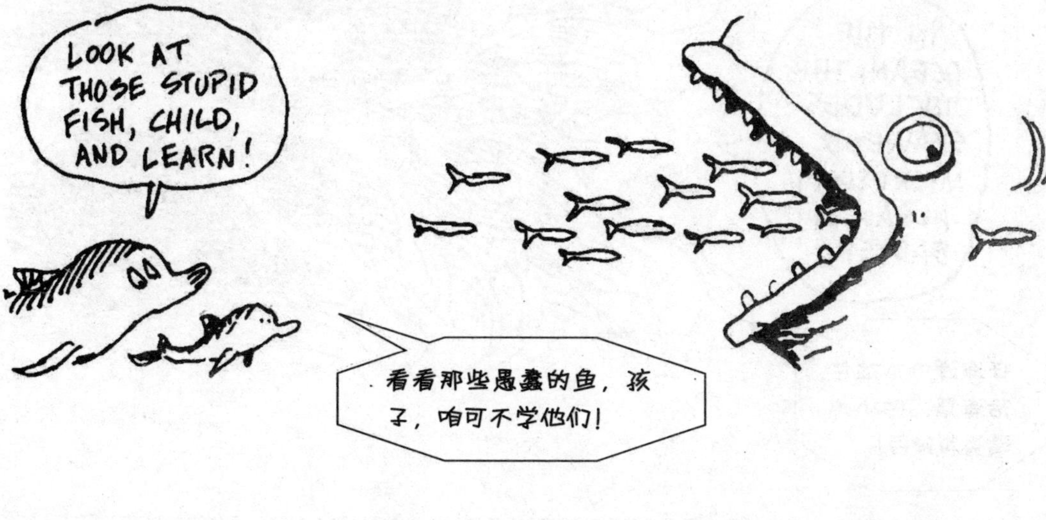

COMPARED WITH LANDLOCKED ECOSYSTEMS, THE OCEANS HAVE MANY LAYERS OF CARNIVORY: PREDATORS PREYING ON PREDATORS PREYING ON PREDATORS. SOME SCIENTISTS BELIEVE THAT THE OCEANS WERE MAXIMALLY HARVESTED FROM WITHIN BEFORE HUMANS STARTED FISHING.

与陆地生态系统相比，海洋生物捕猎具有多层次特征，捕食者捕食的猎物又捕食下一级的猎物。一些科学家认为，在人类开始捕鱼前，海洋生物间的捕获已达到最大限度。

IN THE OCEAN, AS EVERYWHERE, ALL LIFE DEPENDS ON PLANTS, THE PRODUCERS. WE USE THE TERM **NET PRIMARY PRODUCTION** TO DESCRIBE THE TOTAL FOOD MADE AVAILABLE BY PLANTS.

同陆地一样，海洋中的所有生物都依赖于生产者——植物。我们用**净初级生产量**来描述植物制造出的、能够被吸收利用的食物。

TECHNICALLY, NET PRIMARY PRODUCTION (NPP) IS THE DIFFERENCE BETWEEN GROSS PRIMARY PRODUCTION (TOTAL SOLAR ENERGY CONVERTED INTO BIOMASS) MINUS THE ENERGY OF RESPIRATION (THE ENERGY USED BY PLANTS FOR METABOLISM AND GROWTH).

从技术上说，净初级生产量（NPP）等于总初级生产量（指转化成生物量的太阳能总量）减去呼吸消耗的能量（植物用于新陈代谢和成长的能量）。

ENERGY OF RESPIRATION
+
NPP (BIOMASS PRODUCED)
=
TOTAL SOLAR ENERGY USED BY PLANTS

呼吸消耗能量
+
NPP（净生产力）
=
植物吸收的太阳能总量

WITH THAT IDEA IN MIND, LET'S TOUR SOME OF THE OCEAN ZONES...

带着这种概念，我们到一些海里转转吧……

OPEN OCEAN

THE OCEAN FOOD WEB DEPENDS ENTIRELY ON PLANTS IN THE SUNLIT UPPER LAYER (THE ***EUPHOTIC ZONE***).

开放水域

海洋食物网完全依赖于被阳光照射的上层——**透光带**。

大部分植物随水漂流，微小的**浮游植物**为小型**浮游动物**提供养料，而这些浮游动物又是小鱼和大须鲸的食物。小鱼又会成为大鱼、海豹、海豚、鸟类等的食物。死去的生物沉入海底，又会成为蟹类、海胆等食腐生物以及细菌等分解者的营养品。

在海面下200米至500米，是昏暗的**半深海带**，这里是鱿鱼、章鱼、虾和一些凶猛鱼类的家园。至于下面完全黑暗的**深渊带**，我们还不太了解。

DESPITE ALL THIS ACTION, OCEAN LIFE IS SPREAD THINLY OVER A VAST AREA. NPP IN AN AVERAGE PART THE OPEN OCEAN IS VERY LOW, LIKE THAT OF A DESERT. BUT BECAUSE THE OCEAN IS SO LARGE, THE TOTAL PRODUCTION OF FOOD IS IMMENSE.

尽管如此，海洋生物还是稀稀落落地分布在整片海域。平均来讲，净初级生产量在开放水域极低，如同沙漠中荒芜的生命。但海洋如此广阔，食物总产量仍是庞大的。

MOVING IN TOWARD SHORE, WE ENTER A MUCH MORE PRODUCTIVE REGION, WHERE LIFE IS DENSE AND ALL IS SUNNY: THE

COASTAL ZONE.

IT EXTENDS FROM THE OCEAN'S HIGH-TIDE LINE OUT TO THE EDGE OF THE CONTINENTAL SHELF, ABOUT 10% OF THE OCEAN'S AREA.

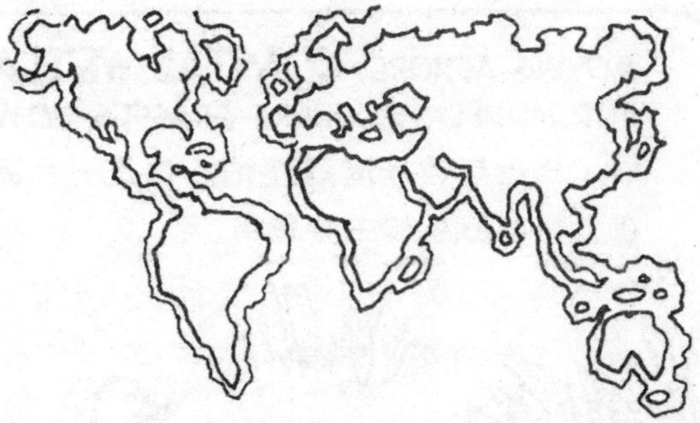

向岸边移动，我们就进入一片更加富饶的区域。这里生物密集、阳光充足，这就是**海岸带**。从潮涨潮落淹没的潮间带一直延伸到大陆架，约占海洋面积的10%。

MANY COASTAL ZONES HAVE ***CORAL REEFS,*** THE OCEAN'S MOST DIVERSE ECO-SYSTEMS. A SINGLE REEF MAY HOUSE 3000 CORAL SPECIES AND COUNTLESS KINDS OF FISH, WORMS, SEAWEED, AND OTHER LIFE FORMS. THE ***PRIMARY PRODUCERS*** HERE ARE ONE-CELLED, GREEN ***DINOFLAGELLATES*** THAT TAKE SHELTER AMONG THE CORAL, USING MINERAL NUTRIENTS SHED BY THEIR HOSTS. THE CORAL RESPONDS BY EATING ORGANIC CARBON COMPOUNDS EXCRETED BY THEIR GUESTS. THE REEF ITSELF IS BUILT UP OF LIMESTONE SHELLS FORMED BY THE LIVING CORAL, AND WHEN THE CORAL DIES, THE SHELLS REMAIN. LIME-STONE (CALCIUM CARBONATE) SEQUESTERS CARBON FROM THE ATMOSPHERE, BUFFERING CLIMATE CHANGE. DESPITE THEIR DIVERSITY, CORAL REEFS ARE SLOW-GROWING AND FRAGILE, AND MANY ARE FALLING VICTIM TO HUMAN ACTIVITIES.

许多海岸带都有**珊瑚礁**，这是海洋中最具多样性的生态系统。一块礁石可能会为3000种珊瑚虫、无数种鱼类、蠕虫、海草和其他生命提供栖身地。这里的**初级生产者**是一种绿色的单细胞**鞭毛藻**，它隐蔽在珊瑚中，吸收珊瑚虫脱落的矿物质元素。珊瑚虫则食用寄居者们排放出的有机碳化合物。礁石本身由活着的珊瑚虫形成的石灰石外壳构成。珊瑚虫死后，外壳存留下来。石灰石（烟酸钙）从大气中吸收碳，从而调节气候变化。尽管珊瑚礁种类很多，但生长缓慢且极其脆弱，许多都沦为人类活动的牺牲品。

MOVING ASHORE, **COASTAL WETLANDS** ARE A MIX OF BAYS, SALT MARSHES, AND MUDFLATS, WHERE GRASSES DOMINATE THE FOOD PRODUCTION.

靠近岸边是**滨海湿地**。这是一片海湾、盐沼、泥滩的混合区域，在这里，禾草在食物制造中起主导作用。

IN WARM CLIMATES, **MANGROVE SWAMPS**—A PARTLY SUBMERGED COLLECTION OF ABOUT 50 SPECIES OF SALT-TOLERANT TREES AND SHRUBS—ARE THE WORLD'S MOST PRODUCTIVE ECOSYSTEM.

红树林沼泽被水浸没，生长着约50种耐盐碱乔木和灌木，在温暖的气候条件下，这里是世界上最富饶的生态系统。

THE MANGROVE'S TALL ROOTS FUNNEL OXYGEN INTO THE MUCK BELOW, AND MANY SEAGOING SPECIES SPEND THEIR YOUTH IN A MANGROVE SWAMP: THERE IS PLENTY TO EAT, AND IT'S EASY TO HIDE IN THE VEGETATION.

红树林的树根很长，能够将氧气传送并渗入淤泥中。许多远洋生物幼小时生活在红树林沼泽，因为这里食物充足，而且在树丛中也容易藏身。

STREAMS

FLOW DOWN FROM THE HILLS WHERE THE WATER, ORIGINATING IN THE OCEAN, FALLS AFTER EVAPORATING.

海水蒸发后，形成降雨，落到山上，形成一条条溪流，潺潺流下。

SWIFT-FLOWING MOUNTAIN STREAMS SUPPORT LITTLE LIFE. PLANTS CLING TO ROCKS, AND FISH ARE COLD-WATER TYPES LIKE TROUT THAT NEED PLENTY OF OXYGEN. (THE STREAMS' TURBULENCE STIRS IN MORE OXYGEN THAN NORMALLY DISSOLVES IN PLACID WATER.)

湍急的山溪中生物很少，植物紧紧地攀附住岩石，而鱼类，如鳟鱼，都属于冷水鱼，需要大量氧气才能生存。（湍急的溪水与普通静水相比，能够蕴含更多的氧气。）

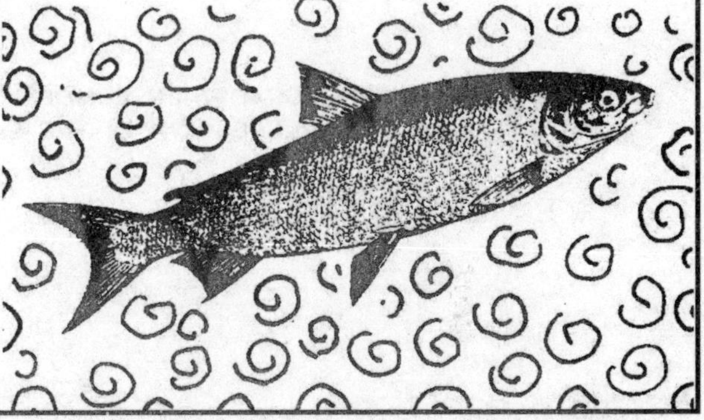

STREAMS MERGE TO FORM WIDER STREAMS ON GENTLER SLOPES. THESE MEANDER THROUGH VALLEYS, SUPPORTING BOTH COLD- AND WARM-WATER FISH WITH LOWER OXYGEN NEEDS.

小溪在坡度较缓处汇集在一起，形成更为广阔的溪流，然后蜿蜒穿过山谷，既供养冷水鱼，又供养暖水鱼，暖水鱼需要的氧气量较少。

WHEN A RIVER IS CHANNELIZED BY HUMANS, CUTTING STRAIGHT ACROSS ONE OF THE MEANDERS, HALF ITS LENGTH MAY BE REMOVED, LEAVING LARGE COMMUNITIES HOMELESS.

当人类开凿渠道，对河曲实行截弯取直时，溪流长度会缩减过半，导致大批水生生物无家可归。

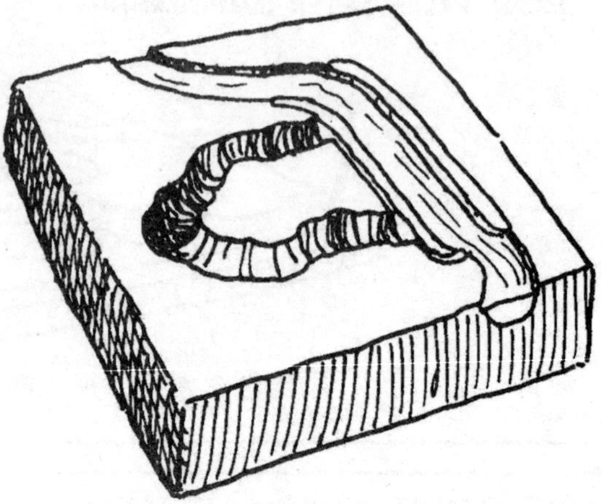

LAKES FORM WHEN RUNOFF OR GROUNDWATER FILLS A DEPRESSION IN THE EARTH. A LAKE HAS FOUR ZONES:

地下水溢出，填满地表洼地，形成**湖泊**。湖泊可分为四个区域：

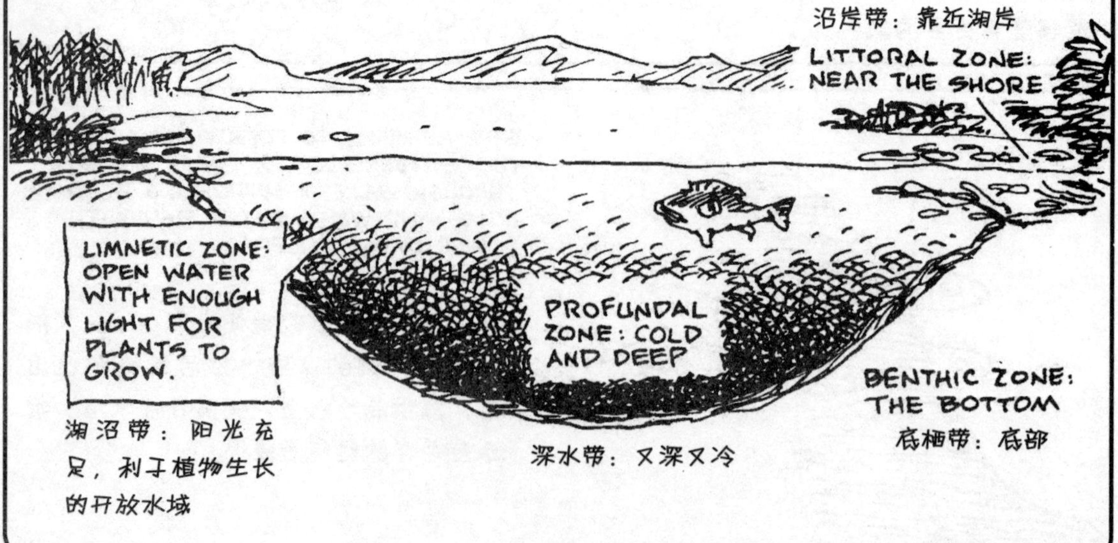

A *EUTROPHIC* ("TRUE FOOD") LAKE IS WELL-NOURISHED, OFTEN SHALLOW, AND MURKY WITH PLANKTON. VARIOUS FISH, SUCH AS BASS, PERCH, SUNFISH, AND PIKE, FLOURISH THERE. THE LITTORAL ZONE IS BROAD, A GOOD HABITAT FOR FROGS AND FISH. OXYGEN IS A COMMON LIMITING FACTOR FOR EUTROPHIC LAKES, AND THE PROFUNDAL ZONE IS OFTEN OXYGEN-FREE.

富营养湖里食物充足，营养丰富，水深较浅，浑浊，充满了浮游生物。这里鱼类丰富，有鲈鱼、河鲈、太阳鱼、梭鱼等。沿岸带比较宽阔，是蛙和鱼类良好的栖息地。对于富营养湖来说，氧气普遍是一个限制因子，深水带常常没有氧。

IN *OLIGOTROPHIC* ("FEW FOOD") LAKES, THERE IS USUALLY A SHORTAGE OF NITRATE AND PHOSPHATE. THE WATER IS CRYSTALLINE AND OFTEN DEEP. AN OLIGOTROPHIC LAKE CAN BE RECOGNIZED BY ITS NARROW SHORE AND CLEAR WATER.

在食物较少的**贫营养**湖中，通常缺乏硝酸盐和磷酸盐。湖水透明且深邃，从狭窄的湖岸和澄清的湖水中便可以辨认出贫营养湖。

INLAND WETLANDS,

LIKE COASTAL WETLANDS, ARE HIGHLY PRODUCTIVE. THESE INCLUDE BOGS, MARSHES, PRAIRIE POTHOLES, FLOODPLAINS, AND WET ARCTIC TUNDRA IN THE SUMMER. WATER IS STILL AND WARM; NUTRIENTS ARE TRAPPED BY PLANT STEMS; PLANKTON ABOUND. MAMMALS, BIRDS, AMPHIBIANS, AND INSECTS ALL FEAST IN WETLANDS.

内陆湿地同沿海湿地一样，生物极其丰富。内陆湿地包括泥沼、草本沼泽、草原壶穴、漫滩以及夏天湿润的极地苔原区。这里水流平静温暖，营养物储存在植物的茎秆中，浮游生物丰富。哺乳动物、鸟类、两栖动物和昆虫都可以在湿地中尽情享受美味。

WETLANDS REGULATE STREAM FLOW BY ACTING LIKE A SPONGE: IN THE DRY SEASON, THEY MAY APPEAR PARCHED AND CRACKED, BUT THEY REMAIN POROUS.

湿地像海绵一样调节着河川径流。在旱季，湿地或许会出现土壤板结、开裂的现象，但仍保持着多孔易渗水的特性。

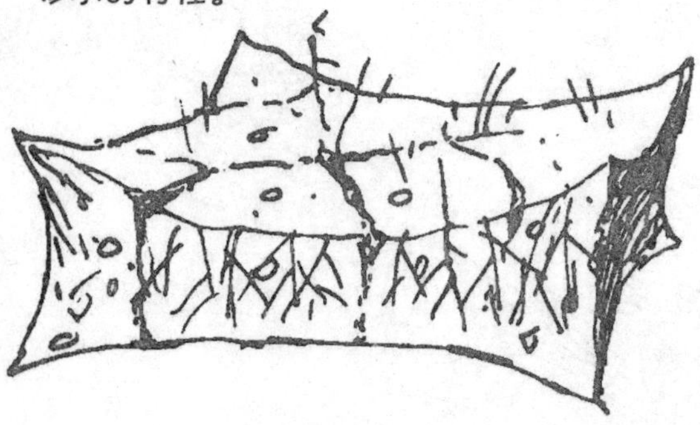

DURING RAIN OR SPRING MELT, THEY SWELL UP, STORING WATER AND RELEASING IT SLOWLY DOWNSTREAM OR ALLOWING IT TO PERCOLATE DOWN INTO THE GROUNDWATER.

在雨季或春天冰雪融化时节，它们就会膨胀并储存大量水分，然后缓慢释放，流向下游或渗入地下水中。

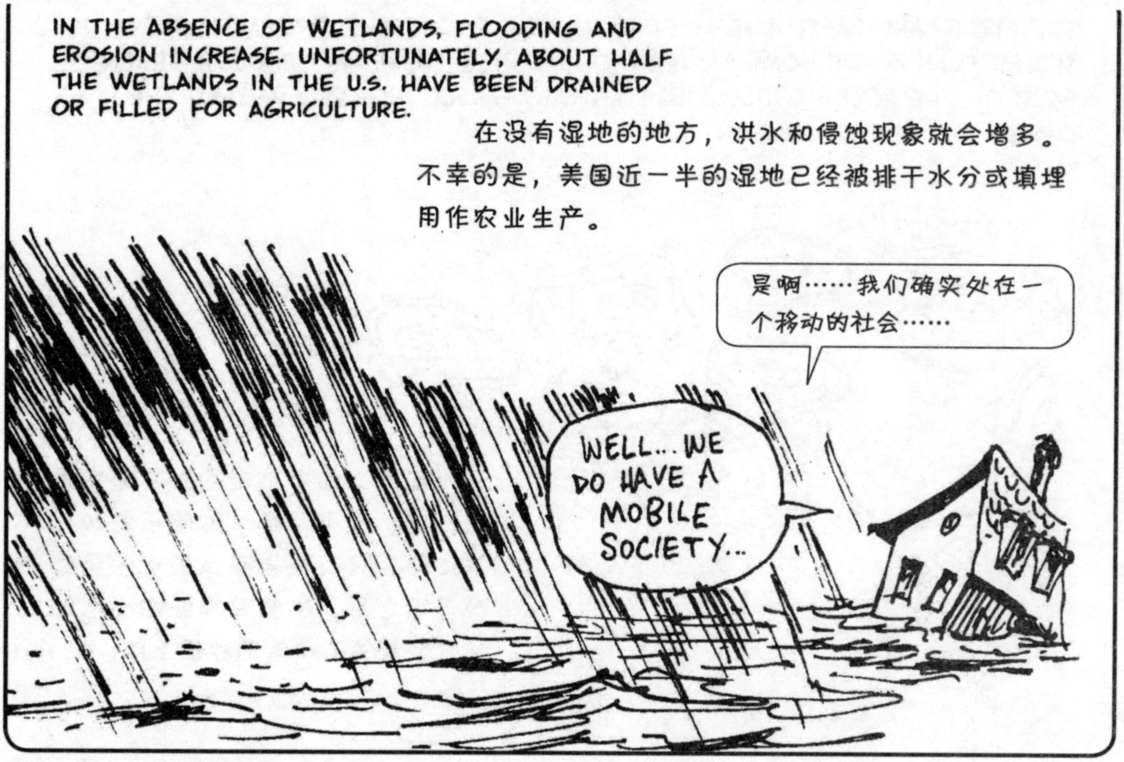

ALTHOUGH LIFE BEGAN IN THE WATER, NOT ALL LIVING COMMUNITIES ARE WATERY. SEVERAL HUNDRED MILLION YEARS AGO, LIFE INVADED THE LAND. DESPITE EVERYTHING YOU HEAR ABOUT LUNGFISH AT THE SCIENCE MUSEUM, THEY COULD NOT POSSIBLY HAVE BEEN THE FIRST LIFE ASHORE—BECAUSE THEY ARE CONSUMERS!

尽管生物起源于水中，但并非所有种群都是水生的。几亿年前，陆地上就有了生物，对此你在自然博物馆听到的可能都是关于肺鱼的讲解，但肺鱼不可能是第一批上岸的生物，因为它们是消费者！

THE FIRST LAND-BASED LIFE MUST HAVE BEEN PRODUCERS, PROBABLY ONE-CELLED PLANTS AND SOME BACTERIA. ONLY AFTER THERE WAS ENOUGH GREEN SCUM ON THE ROCKS COULD SMALL ANIMALS—ALSO ONE-CELLED—LEAD THE CHARGE OF THE CONSUMERS.

陆地上最早出现的生物必定是生产者，很可能是单细胞植物和一些细菌。只有在岩石上出现大量绿色浮垢之后，那些单细胞的小型动物才会带领消费者们冲锋上阵。

SINCE THEN, THESE EARLY ALGAE, LICHEN, MOSSES, AND FERNS, ALONG WITH THEIR CONSUMERS, THE INSECTS, ARTHROPODS, AND FISH, HAVE EVOLVED INTO MILLIONS OF DIFFERENT SPECIES, WHOSE COMMUNITIES ARE THE SUBJECT OF OUR NEXT CHAPTER...

从那之后，早期的藻类、地衣、藓类植物、蕨类植物同它们的消费者——昆虫、节肢动物、鱼类等一起，进化成了数百万种不同物种。这些生物便是我们下一章的主题……

CHAPTER 5

...AND DRY

◆第五章◆

陆生生物

ON LAND, BASIC LIFE COMMUNITIES ARE CALLED ***BIOMES***. BIOMES DEPEND ON CLIMATE AND GEOGRAPHY: WHERE CLIMATE IS DIFFERENT, SO ARE THE PLANTS AND ANIMALS, WHILE SIMILAR BIOMES MAY EXIST IN WIDELY SEPARATED AREAS THAT HAVE SIMILAR ENVIRONMENTAL CONDITIONS.

在陆地上，生物聚居的基本群落被称为**生物群系**，生物群系对气候和地理条件具有依赖性。也就是说，在不同的气候环境里生长着不同的植物和动物。具有相似特点的生物群系会广泛分布在各个地区，这些地区的环境条件也具有相似的特点。

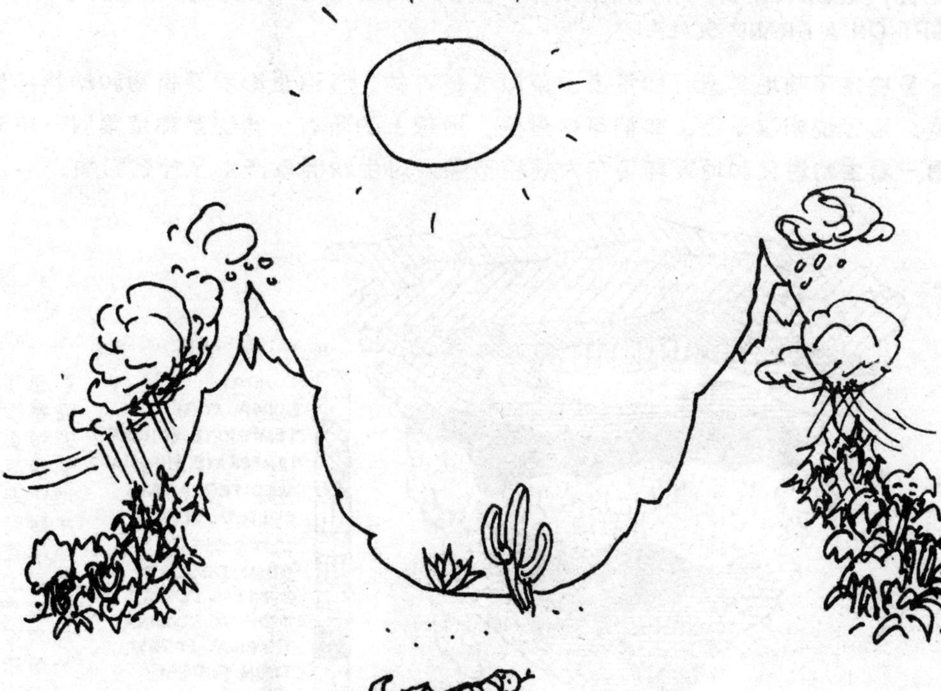

AS A GENERAL RULE, CLIMATE COOLS AS YOU HEAD AWAY FROM THE EQUATOR AND TOWARD THE POLES... AND ALSO AS YOU RISE ABOVE SEA LEVEL. RAINFALL IS HEAVY AT THE EQUATOR, LIGHTER AT THE TROPICS OF CANCER AND CAPRICORN, AND RISES AGAIN POLEWARDS.

总的来说，离赤道越远、离两极越近的地区，气候越凉爽……海平面越高，气温越低。赤道地区的降雨量最多，南北回归线之间的地区降雨较少，靠近两极的地区降雨量又增多。

PRECIPITATION DETERMINES WHETHER THE LAND IS DESERT, GRASSLAND, OR FOREST, AND TEMPERATURE ALSO INFLUENCES THE STRATEGIES OF BOTH PLANTS AND ANIMALS. WE CAN SUMMARIZE THIS BY LUMPING ALL THE WORLD'S BIOMES INTO A SUPERCONTINENT, SHOWING WHAT EVOLUTION AND GEOGRAPHY HAVE WROUGHT ON A GRAND SCALE.

降水量决定了陆地类型，即荒漠、草原或是森林。气温也影响着植物和动物的生长和繁殖方式。为了说明这一点，我们可以设想，地球上的所有生物群系都聚集到一块超级大陆上，看一看生物进化和地理环境在大规模范围内对生物群系产生了什么影响。

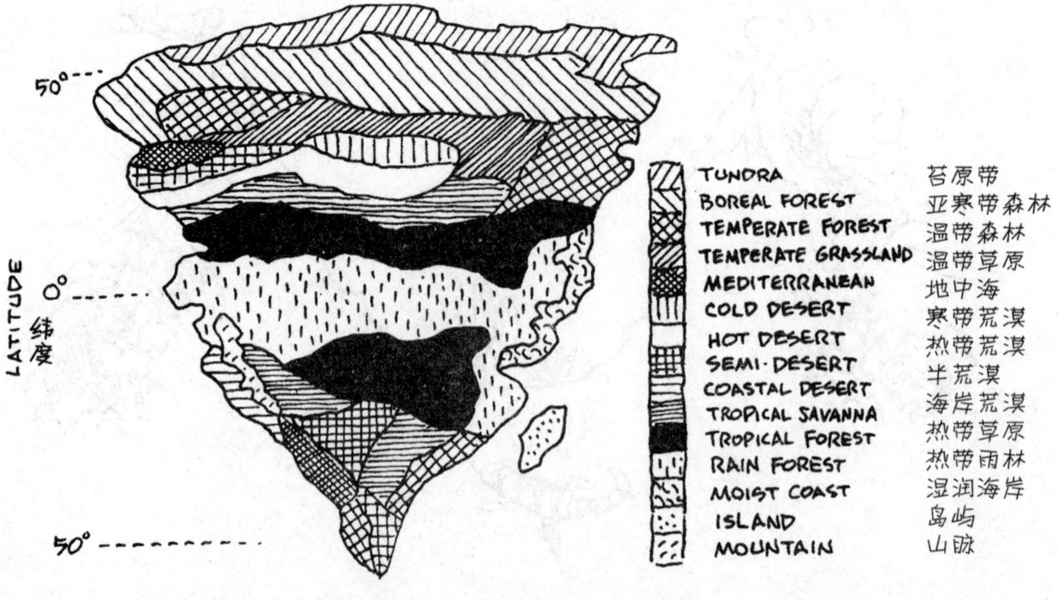

HERE IS A TOUR OF THE WORLD'S MAJOR BIOMES:

POLAR GRASSLAND

(OR ARCTIC TUNDRA)

让我们参观一下地球上的主要生物群系：

IS AN ICY, TREELESS BIOME NORTH OF THE TREELINE AND BELOW THE ARCTIC CIRCLE (AND JUST ABOVE THE TIMBERLINE ON MOUNTAINS). THE TUNDRA'S LIMITING FACTORS ARE ***HEAT*** AND ***LIGHT*** (OR RATHER, COLD AND DARKNESS!), SO FEW PRODUCER SPECIES CAN LIVE HERE. THE SOIL IS PERMANENTLY FROZEN, ICING OUT TREE ROOTS. WINTERS ARE FIERCE.

极地草原或称“极地苔原带”，是一片冰冷、没有树木的地带，位于森林线以北、北极圈以南（高山上就是森林线以上的地带）。苔原带的限制因子是**热**和**光**（更准确地说是寒冷和黑暗！），所以这里很少有生产者物种生存。土壤是永冻土，树根不能生长，冬季气候恶劣。

DURING THE BRIEF SUMMER THAW, SURFACE SOIL BECOMES A QUAGMIRE OF PUDDLES AND BOGS; LOW-GROWING PLANTS BUD AND BLOSSOM. HORDES OF INSECTS HATCH; BIRDS ARRIVE TO EAT THE BUGS. THE PERMANENT RESIDENTS ARE MOSTLY BURROWING PLANT-EATERS THAT HUDDLE UNDER THE SNOW IN WINTER. BIG HERBIVORES VISIT IN THE SUMMER AND THEN HEAD SOUTH. FOX, LYNX, AND BEAR LIVE HERE TOO.

在短暂的夏季永冻土融化时期，表层土壤就变成一个个坑坑洼洼的泥潭、泥沼。在这里，低矮的植物萌芽开花，成群的昆虫孵化成长，鸟儿停留在沼泽里寻觅食物。这里的永久居民主要是穴居的植食动物，冬天，它们便蜷缩在雪地下；夏天，个头较大的食草动物也会到来，然再返回南方。在夏季，狐狸、猞猁、熊都会来到这里。

缓慢生长的植物和薄薄的土层，使苔原带成为最脆弱的生物群系：100年前留下的车辙至今依稀可辨。

THE SLOW-GROWING PLANTS AND SHALLOW SOIL MAKE TUNDRA THE MOST FRAGILE BIOME: 100-YEAR-OLD WAGON TRACKS ARE STILL VISIBLE TODAY.

BOREAL FOREST
(OR "TAIGA")

COLD AND WET, THE TAIGA STRETCHES ACROSS NORTH AMERICA AND EURASIA, COVERING 11% OF ALL LAND. WINTERS ARE LONG AND COLD, BUT SUMMERS ARE LONGER AND WARMER THAN IN THE TUNDRA. BOREAL FORESTS INCLUDE YOSEMITE, SEQUOIA, AND YELLOWSTONE NATIONAL PARKS.

亚寒带森林带或称“泰加林带”。冰冷而又潮湿，横跨北美和亚欧大陆，占大陆总面积的11%。这里冬季漫长而寒冷，但夏季比苔原带更长，更加温暖。亚寒带森林带包括位于加利福尼亚州中部的约塞米蒂国家公园、西北太平洋沿岸的红杉国家公园和美国中西部的黄石国家公园。

DOMINATED BY CONIFEROUS (CONE-BEARING) EVERGREEN TREES, THE RELATIVELY INFERTILE FOREST FLOOR IS STREWN WITH NEEDLES AND LEAF LITTER, INTERRUPTED BY A FEW HARDY SHRUBS AND PLANTS. LARGE HERBIVORES INCLUDE MOOSE, MULE DEER, CARIBOU, AND ELK; SMALL ONES INCLUDE HARE, SQUIRRELS, AND OTHER RODENTS. INSECT DIVERSITY IS LOW, WITH BUTTERFLIES, BEETLES, WASPS, AND FLIES. MAJOR PREDATORS INCLUDE THE TIMBERWOLF, LYNX, FOX, MARTEN, WOLVERINE, MINK, OTTER, ERMINE, AND WEASEL.

这里的植被以结球果的常绿树为主，森林土壤相对贫瘠，铺满了松针和落叶，间杂着少量耐寒的灌木和植物。大型食草动物有驼鹿、骡鹿、驯鹿和马鹿，小型食草动物有野兔、松鼠和其他啮齿类动物。昆虫种类很少，只有蝴蝶、甲虫、黄蜂和蝇类。主要的食肉动物有森林狼、猞猁、狐狸、貂、貂熊、水貂、水獭、白鼬和其他鼬类。

TAIGA IS THREATENED BY HEAVY LOGGING, SINCE CONIFEROUS TREES GROW RAPIDLY, GIVING THEM HIGH ECONOMIC VALUE FOR PULP AND LUMBER PRODUCTS.

针叶树生长速度快，用于造纸和木制产品生产可以带来很大的经济效益，因此，泰加林正面临着过度砍伐的威胁。

TEMPERATE DECIDUOUS FOREST

ONCE COVERED MOST OF CENTRAL EUROPE, EASTERN CHINA, AND EASTERN NORTH AMERICA... BUT MOST OF IT HAS BEEN CLEARED FOR FARMS.

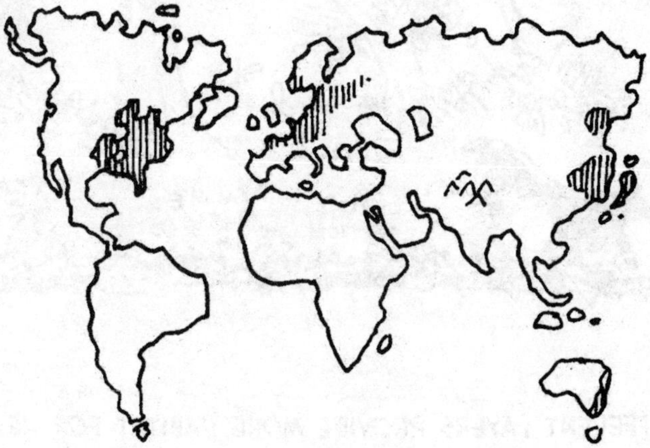

温带落叶林带曾经覆盖了欧洲中部、中国东部和北美东部大片区域，但现在大多都已被开垦为农田。

HERE RAIN IS ABUNDANT, MODERATE TEMPERATURES CHANGE SEASONALLY, AND THE GROWING SEASON IS 4-6 MONTHS.

这里降水充沛，气候温和，四季分明。生长期可达4~6个月。

THE DECIDUOUS FOREST IS STRATIFIED, OR LAYERED, WITH A HIGH CANOPY, AN UNDERSTORY OF SMALLER HARDWOODS, A SHRUB LAYER, THEN GROUND-COVERING HERBS, AND A FINAL LAYER OF MOSS AND LICHEN. AS TEMPERATURES DROP IN THE FALL, SO DO THE LEAVES, DECOMPOSING TO PRODUCE RICH SOIL THAT SUPPORTS A WIDE VARIETY OF PLANTS.

落叶林层次分明，上面是高高的树冠，向下依次是小型阔叶树、灌木、近地面草本植物，最下面是苔藓和地衣层片。当秋季气温下降时，树叶飘落、分解，为土壤提供养分，供多种植物生长。

DIFFERENT LAYERS PROVIDE MORE HABITAT FOR HERBIVORES (WHITETAIL DEER, BLACK BEAR, BEAVER, PORCUPINE, OPOSSUM, RACCOON, SKUNK, CHIPMUNK, GRAY SQUIRREL, SHREW, AND THE COTTONTAIL RABBIT), SO THERE ARE MORE CARNIVORES (WOLVES, BOBCATS, GRAY FOXES, MOUNTAIN LIONS). THIS BIOME HAS BEEN THE MOST REDUCED BY HUMAN ACTIVITY, AND RELATIVELY FEW VIRGIN STANDS REMAIN.

多层次落叶林为许多动物提供了栖息地，食草动物有白尾鹿、黑熊、海狸、豪猪、负鼠、浣熊、臭鼬、花栗鼠、灰松鼠、鼩鼱和棉尾兔等。食肉动物就更多了，如狼、短尾猫、灰狐和美洲狮等。然而人类活动已经致使落叶林带面积锐减，原始森林几乎没有了。

GRASSLAND

IS FOUND WHERE RAINFALL IS TOO LOW FOR FORESTS BUT HIGH ENOUGH TO KEEP DESERTS FROM FORMING. WINTERS ARE COLD, SUMMERS HOT AND DRY, AND THE WIND BLOWS ALMOST ALL THE TIME. THE MOISTER THE GRASSLAND, THE TALLER THE GRASS.

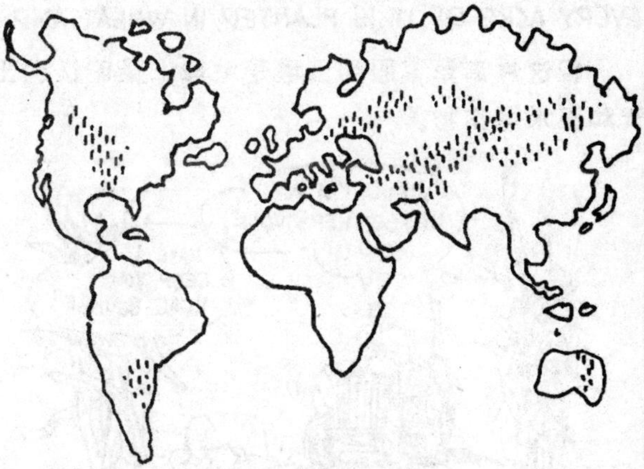

草原带降水量相对较少，不足以供养森林生长，但又足以避免荒漠形成。这里冬季寒冷，夏季炎热干燥，一年中的大部分时间都是大风天气。气候越湿润，禾草长得越高。

BEFORE AGRICULTURE, THE AMERICAN GREAT PLAINS WERE DOMINATED BY LARGE HERBIVORES: BISON, ANTELOPE, ELK; SMALL HERBIVORES, LIKE PRAIRIE DOGS AND JACKRABBIT; AND PREDATORS, INCLUDING WOLF, COYOTE, PANTHER, AND PEOPLE.

在农耕前，美国大平原上生活着各种动物，其中大型食草动物有野牛、羚羊、马鹿，小型食草动物有土拨鼠、长耳大野兔；食肉动物则有狼、郊狼和美洲豹，当然也有人类。

BUT THIS TALLGRASS SOIL IS AMONG THE WORLD'S RICHEST, AND NOW NEARLY EVERY ACRE OF IT IS PLANTED IN WHEAT AND CORN.

但这片高草草原的土壤是地球上最肥沃的土壤之一，现在几乎每英亩土地都种上了小麦和玉米等作物。

GRASSLANDS ARE UNDER CONSTANT STRESS FROM WIND, SUN, AND TEMPERATURE EXTREMES. DROUGHT OR MISUSE CAN LEAD TO WIND EROSION, STRIPPING IRREPLACEABLE TOPSOIL AND TURNING GRASSLAND INTO BARREN DESERT.

草原带正不断承受着来自大风、太阳和极端气温的威胁，干旱或是土地资源滥用都会导致风蚀，大风会吹走不可填补的表层土壤，把草原变成光秃秃的荒漠。

DESERTS

COVER ALMOST 1/4 OF THE LAND'S SURFACE. ***COLD DESERTS,*** LIKE THE GOBI, HAVE WARM SUMMERS AND FRIGID WINTERS. ***TEMPERATE DESERTS,*** LIKE THE MOJAVE IN SOUTHERN CALIFORNIA, HAVE HOT SUMMERS AND COOL WINTERS. ***TROPICAL DESERTS,*** LIKE THE SOUTHERN SAHARA, ARE JUST PLAIN HOT AND SUPPORT LITTLE LIFE. ALL DESERTS HAVE LITTLE WATER AND LOW SPECIES DIVERSITY. DESERT PLANTS, WHICH HAVE EVOLVED TO COLLECT AND CONSERVE PRECIOUS WATER, GROW SLOWLY, SO DESERT ECOSYSTEMS ARE FRAGILE.

荒漠带约占陆地表面积的1/4。**冷荒漠**（如戈壁）夏季温暖，冬季则非常寒冷。**温性荒漠**（如加州南部的莫哈韦沙漠）夏季炎热，冬季则很凉快。**热带荒漠**（如南撒哈拉大沙漠）一年四季炎热无比，几乎没有生物生长。所有的荒漠都严重缺水，仅有少量物种生存。荒漠植物经过长期进化，学会了收集并储存珍贵的水，但生长极其缓慢，因此荒漠生态系统非常脆弱。

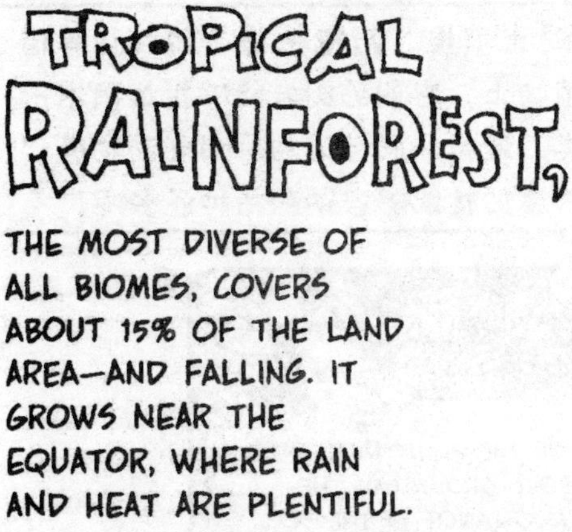

热带雨林带最具多样性，约占陆地面积的15%，只是这个比例正在下降。雨林主要生长在赤道附近，这里气候温暖，降雨丰富。

热带雨林带的限制因子是光照，因为浓厚的叶冠遮挡住下面的森林，空气静止，植物不能依赖风传粉，但是热带花朵与鸟类、昆虫和蝙蝠协同进化，依靠它们来授粉。

除了空旷地带的边缘外，地面几乎没有任何植物。无论什么落到地面上，都会被成群的昆虫迅速吃掉。

植物养分大多储存在地面上方的树木中，而不是土壤里。**附生植物**（如兰花等借势长到空中的植物），紧紧攀附在树木和其他藤本植物上，它们的树叶能够拦截雨水形成水塘，内部形成自身的微小生态系统。昆虫、鸟类、蛙类、爬行动物、哺乳动物等，成群结队地挤到枝叶上。热带雨林的一棵树比一整片泰加林能够容纳的物种还多呢！

热带雨林带中的一些地方，土壤非常贫瘠，含有**红土**。当人们滥伐森林、使土壤完全暴露在太阳下的时候，土壤中的铝铁化合物在太阳的烘烤下会变得像混凝土一样。

TROPICAL GRASSLAND,

OR ***SAVANNA***, COVERS 11% OF THE LAND. IT OCCURS IN AREAS WITH TWO LONG DRY SEASONS, NO WINTER, AND ABUNDANT RAIN THE REST OF THE YEAR.

热带草原带或称**稀树草原带**，占陆地面积的11%。这样的地区有两个长长的旱季，没有冬季，其余的时间降雨充沛。

THE SAVANNA HAS BROAD EXPANSES OF TALL GRASSES AND SCATTERED TREES. TROPICAL SAVANNAS GROW QUICKLY, CREATING AS MUCH PLANT MATTER AS A TROPICAL RAINFOREST, FOOD FOR HUGE HERDS OF HERBIVORES.

稀树草原带上是辽阔的高秆禾草，树木稀稀落落。这些热带植物生长速度快，这里的植物生物量和热带雨林一样多，并可为大量的食草动物提供食物。

BETWEEN ANY TWO BIOMES LIES A TRANSITION ZONE CALLED AN **ECOTONE.** THIS REFERS TO ANY REGION WHERE TWO BIOMES OR ECOSYSTEMS MEET. AN ECOTONE CONTAINS SPECIES FROM BOTH COMMUNITIES AND MAY HAVE SOME OTHER CHARACTERISTIC SPECIES ALL ITS OWN.

任何两个生物群系之间的过渡区域称为**生态过渡带**，也就是两个生物群系或生态系统相连的区域。生态过渡带中含有两个生物群中的物种，还可能有其他特有的物种。

THE INCREASED VARIETY AND DENSITY OF SPECIES IN AN ECOTONE IS KNOWN AS THE **EDGE EFFECT.**

*译注：原文为双关语（on edge 意为“烦躁的”），故译文中也宜含有“边缘”二字。

生态过渡带中物种种类和数量增多的现象被称为**边缘效应**。

AND JUST TO MAKE LIFE EVEN MORE INTERESTING, THE EDGES HAVE A WAY OF MOVING AROUND: NOTHING STAYS PUT IN NATURE!

更有趣的是，这些边缘不断发生的变化使得生物也不断变化：自然界中没有静止不变的东西嘛！

ALTHOUGH WE'VE DESCRIBED OUR BIOMES AS IF THEY WERE STABLE, IN FACT, THEY'RE ALWAYS CHANGING. SPECIES COME AND GO... THE ENVIRONMENT CHANGES... BORDERS SHIFT... NEW OPPORTUNITIES AND CHALLENGES ARISE...

虽然在描述这些生物群系时仿佛它们都是静止的，可事实上，生物群系一直在不断变化。旧的物种消失，新的物种出现……环境在变化……边界在移动……新的机遇和挑战也在不断产生……

IN PARTICULAR, LIFE CAN COLONIZE VIRGIN TERRITORY AND TURN BARREN LAND INTO A THRIVING ECOSYSTEM, A PROCESS KNOWN AS

ECOLOGICAL SUCCESSION

(OR COMMUNITY DEVELOPMENT).

尤其重要的是，生物能够开拓处女地，把荒芜的土地变成繁荣的生态系统，这个过程便是**生态演替**(又叫群落发展）。

PRIMARY SUCCESSION

OCCURS WHEN LIFE INVADES AN AREA WITH NO TRUE SOIL. FOR EXAMPLE:

原生演替是在生物进入一片完全没有植被的原生裸地后发生的演替，这些地方没有真正意义上的土壤，例如：

AT FIRST, A FEW HARDY MICROBES, MOSSES, AND LICHENS INVADE THE AREA.

最初，少量顽强的微生物、苔藓和地衣开始在这种地方生长。

THESE PIONEER SPECIES ARE MOSTLY SMALL PRODUCERS, PLUS A FEW DECOMPOSERS. PIONEER PLANTS ARE SMALL, LOW-GROWING ANNUALS, r-STRATEGISTS THAT SPEND THEIR ENERGY MAKING LOTS OF SEEDS RATHER THAN DEVELOPING MUCH IN THE WAY OF ROOTS, STEMS, OR LEAVES.

这些先驱物种大多是小型生产者，其中有少数分解者。先驱植物是矮小的一年生植物，采用大量繁殖策略。它们全力以赴产出大量种子，而不是着力使自己生长根、茎或叶。

EVENTUALLY, SOIL IS CREATED, WEEDS SPROUT, AND ANIMALS ARRIVE, SEEKING FOOD OR COVER.

就这样，土壤终于慢慢形成了，杂草开始丛生，动物们也来到这里寻找食物或藏身之处。

SECONDARY SUCCESSION

BEGINS WHERE VEGETATION HAS BEEN REMOVED BUT SOIL IS INTACT. FOR EXAMPLE:

次生演替发生在那些植被已被破坏但土壤仍然完好的区域，例如：

ABANDONED FARMLAND
被弃耕的农田

BURNED OR LOGGED FOREST
被烧毁或砍伐的森林

LAND RECLAIMED FROM FLOOD.
洪水过后复垦的土地

HERE NEW PLANTS CAN SPROUT WITHIN WEEKS. FIRST COME THE ANNUAL WEEDS...

这里的新生植物可以在几周内发芽，最先出现的是一年生杂草……

THEN THE GRASSES, SHRUBS, A YOUNG (SOFTWOOD) FOREST, AND FINALLY A MATURE, HARDWOOD FOREST.

然后长出禾草、灌木、小针叶林，最终是成熟的阔叶林。

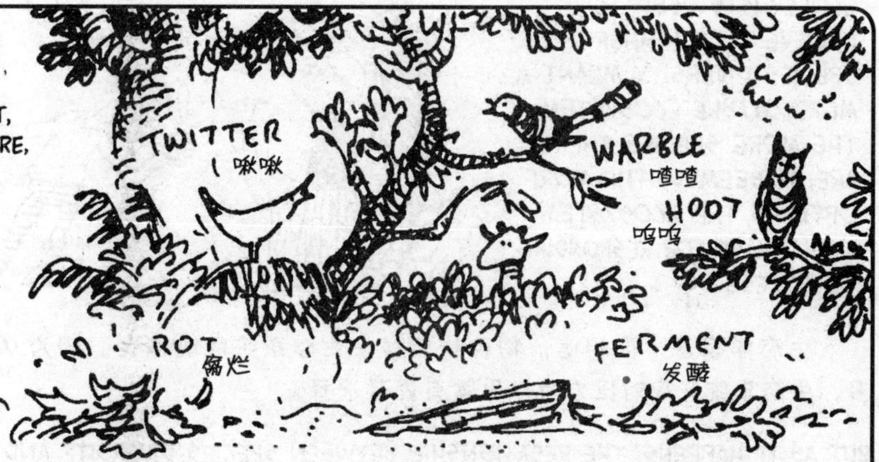

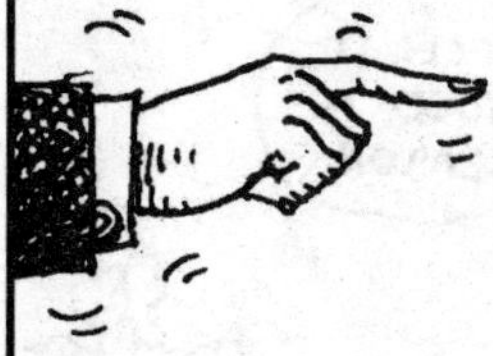

NOTE: ALTHOUGH WE HAVE MADE SUCCESSION SOUND PREDICTABLE AND ORDERLY, THE EXACT SEQUENCE OF SPECIES AND COMMUNITY TYPES CAN BE EXTREMELY VARIABLE, DEPENDING ON LOCAL CONDITIONS.

请注意：尽管我们在描述演替过程时听起来好像演替过程是可以预测的，而且是井然有序的，但事实上，不同物种和不同群落类型的演替过程因当地条件不同是非常复杂多样的。

BY THE TIME AN ECOSYSTEM MATURES, IT HAS DIVERSE SPECIES, STABLE POPULATIONS, AND COMPLEX INTERDEPENDENCIES AMONG ITS PLANTS AND ANIMALS. TO SURVIVE, THE RESIDENTS NEED LARGE AREAS OF UNDISTURBED WILDERNESS.

生态系统形成后，物种就会多样，数量就会稳定，植物和动物间保持着错综复杂的相互依赖关系。为了生存，动植物都需要大片未受破坏的野地。

ECOLOGISTS USED TO BELIEVE THAT HIGHER SPECIES DIVERSITY MEANT A MORE STABLE ECOSYSTEM. THE MORE SPECIES THERE ARE, IT SEEMED, THE MORE OPTIONS THE ECOSYSTEM WOULD HAVE FOR RESPONDING TO STRESS.

生态学家曾一度相信，物种越多样，生态系统就越稳定。因为从表面上看，物种越多，生态系统在应对压力时似乎就有更多选择。

BUT AS IT HAPPENS, THE RELATIONSHIP BETWEEN SPECIES DIVERSITY AND STABILITY CAN BE TENUOUS, EVEN NONEXISTENT. COMPLEX ECOSYSTEMS CAN BE FRAGILE!!

但事实证明，物种多样性和稳定性之间的相关性可能很小，甚至可能根本就没有相关性。错综复杂的生态系统可能是很不稳定的！

THIS IS BECAUSE ECOLOGICAL STABILITY ITSELF IS A COMPLEX PHENOMENON, WITH AT LEAST THREE DIFFERENT ASPECTS:

INERTIA, CONSTANCY, AND RESILIENCE.

这是因为生态稳定性自身便是一个复杂的现象，至少有三个不同方面的原因：**惯性**、**恒定性**和**恢复力**。

INERTIA IS THE ABILITY OF AN ECO-SYSTEM TO RESIST CHANGE. A RAIN FOREST, FOR EXAMPLE, HAS HIGH INERTIA.

惯性是生态系统抵抗变化的能力。例如，雨林的惯性就很大。

CONSTANCY IS THE ABILITY OF A LIVING SYSTEM, LIKE A POPULATION, TO PRESERVE ITS NUMBERS. THE HUMAN RACE HAS TERRIFIC CONSTANCY.

恒定性是生态系统维护种群数量、结构稳定性的能力。例如，人种就具有极强的恒定性。

无论是瘟疫、饥荒还是战争，都不影响我们的发展！

RESILIENCE IS A SYSTEM'S ABILITY TO RESTORE ITSELF AFTER SUFFERING AN OUTSIDE DISTURBANCE. GRASSLANDS, FOR EXAMPLE, SPRING BACK AFTER A FIRE, BECAUSE SO MUCH OF THE PLANT MATTER IS SUBSURFACE ROOTS.

恢复力是生态系统在遭受外力破坏后重生的能力。例如，草在火灾后会重生，这是因为草的地下根异常繁盛。

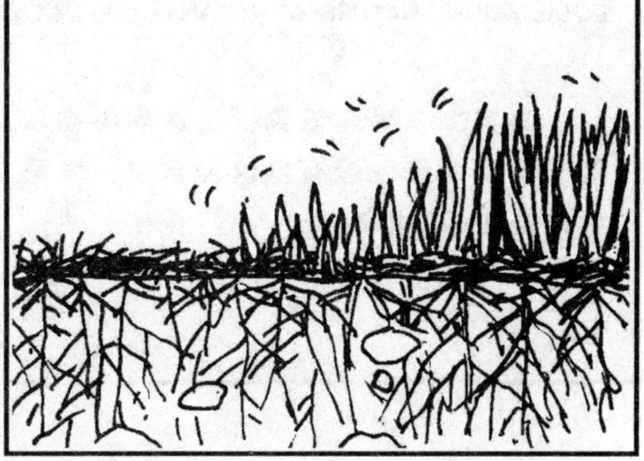

BUT THE DIVERSE, HIGH-INERTIA RAIN FOREST HAS VERY LOW RESILIENCE: ONCE CLEARED, THE FOREST IS GONE, BECAUSE SOIL NUTRIENTS AND THE WATER CYCLE CAN NO LONGER SUPPORT IT.

但是，雨林虽物种丰富，惯性很大，但恢复力却很小。雨林一旦被砍伐，森林就会消失，因为土壤养分和水循环都不足以支持整个系统重生。

BY CONTRAST, GRASSLAND HAS LOW SPECIES DIVERSITY AND LOW INERTIA, BUT HIGH RESILIENCE.

相比之下，草原的物种比较贫乏，惯性较小，却有很高的恢复力。

ANOTHER MISCONCEPTION IS THAT ECOSYSTEMS HAVE A FAIRLY STABLE ***EQUILIBRIUM*** STATE, NEAR WHICH THEY REMAIN.

还有一种错误观点是，生态系统存在一个相当稳定的**平衡**状态，也一直保持着一种近乎平衡的状态。

IN FACT, ECOSYSTEMS ARE RARELY, IF EVER, AT EQUILIBRIUM. NATURE CONSTANTLY FLUCTUATES.

事实上，生态系统很少达到平衡状态，可能从来就没有过这种状态。大自然总是在不停地波动。

NATURAL SYSTEMS ARE IMMENSELY COMPLEX, WHILE OUR UNDERSTANDING OF THEM REMAINS LIMITED. TO SOME EXTENT, IT'S BOUND TO STAY THAT WAY: WE CAN RARELY DO CONTROLLED EXPERIMENTS ON WHOLE ECOSYSTEMS (THAT WOULD REQUIRE A SECOND, IDENTICAL ECOSYSTEM!), AND WE CANNOT KNOW OR OBSERVE ALL THE RELEVANT VARIABLES.

自然系统极其复杂，而我们的理解还十分有限。但从某种程度上讲，我们也只能这样不了了之。这是因为我们无法对整个生态系统做对照试验（对照试验还需要另一个相同的生态系统！），而且我们也不可能知道或是观察到所有的相关变量。

SO WE EXPERIMENT ON SIMPLE SYSTEMS IN THE LAB, OR SIMULATE THEM ON THE COMPUTER. THE RESULTS ARE OFTEN SUGGESTIVE, BUT WE HAVE LIMITED CONFIDENCE THAT THEY CAN BE APPLIED TO THE REAL WORLD.

所以，我们只能在实验室里做些简单的实验，或是用计算机进行模拟实验。实验结果通常具有启发性，但并不确定它们是否就可以真正用于现实世界。

NEVERTHELESS, WE DO KNOW A FEW THINGS ABOUT HOW SPECIES INTERACT... AND IN OUR NEXT CHAPTER WE TURN TO ONE OF THE SIMPLEST INTERACTIONS OF ALL...

尽管如此，关于物种间的相互作用我们还是知道一些的……下一章我们就来了解一下物种之间最简单的关系……

· CHAPTER 6 ·

◆第六章◆

ALTHOUGH YOU MAY NOT USUALLY THINK OF IT THIS WAY, THE ACT OF EATING A FELLOW CREATURE IS A WAY OF CAPTURING ITS ***CHEMICAL ENERGY.***

尽管通常你可能不会这么想，但是进食其他生物的行为却是捕获**化学能**的一种方式。

WHEN THE VEGGIE BURGER HITS YOUR STOMACH, ENZYMES GO TO WORK BREAKING DOWN ITS CHEMICAL CONSTITUENTS AND DEPLOYING THEM TO VARIOUS SYSTEMS OF YOUR BODY.

蔬菜汉堡一进入胃，各种酶便开始发挥作用，将汉堡的化学成分分解并输送到身体的各种系统中。

THE BURGER'S FATS, FOR EXAMPLE, ARE ***BURNED*** IN YOUR CELLS, COMBINING WITH THE OXYGEN YOU INHALE TO RELEASE ***HEAT ENERGY,*** JUST LIKE A REGULAR FIRE, ONLY SLOWER.

例如，汉堡的脂肪与吸入的氧气结合，在体内细胞中**燃烧**并释放**热能**，这个过程就像正常情况下燃烧的火，只是速度更慢。

YOU USE THAT ENERGY TO WARM YOUR BODY TO 37° C, TO MOVE AROUND, TO THINK, TO BREATHE, IN SHORT, TO LIVE!

这种能量使你的体温达到37℃，能支持你的活动，使你能够思考和呼吸，简单地说，就是生存。

THE FOOD ALSO SERVES AS YOUR BODY'S SOURCE OF RAW MATERIALS.

食物同时也是身体的原料来源。

HOW THE ENERGY FLOWS THROUGH AN ECOSYSTEM DEPENDS ON ***WHO EATS WHOM.*** THE SEQUENCE OF DINERS AND DINNERS IS CALLED THE ***FOOD CHAIN,*** OR MORE ACCURATELY THE ***FOOD WEB,*** BECAUSE THE CHAINS ARE INTERLINKED AND MAY SHIFT AROUND.

能量在生态系统中的流动方式取决于**谁吃谁**。吃与被吃的关系链称为**食物链**，更确切地说是**食物网**，因为这一食物链是相互联接，并有可能移位的。

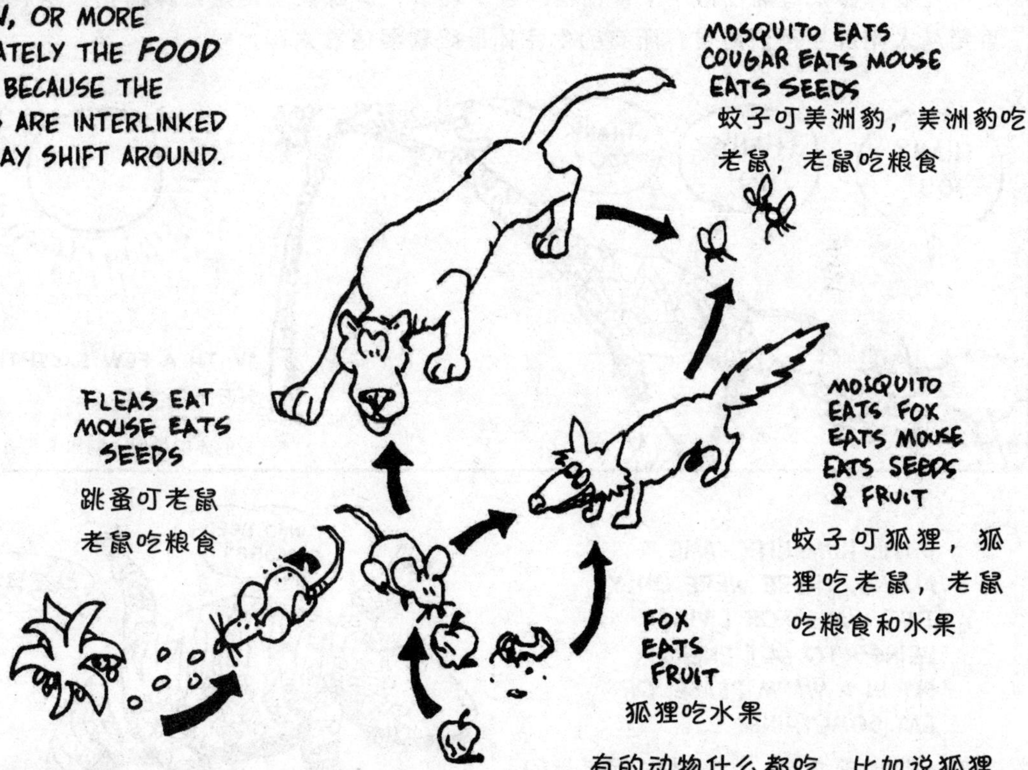

A GENERALIST LIKE THE FOX, FOR EXAMPLE, EATS WHATEVER IS ABUNDANT AND IN SEASON: MICE, BERRIES, GRASSHOPPERS, FALLEN APPLES...

有的动物什么都吃，比如说狐狸，它会吃任何丰富而应季的食物：老鼠、浆果、蚂蚱及树上落下的苹果……

ACCORDING TO THE

FIRST LAW OF THERMODYNAMICS,

ENERGY IS NEITHER CREATED NOR DESTROYED.

根据**热力学第一定律**，能量既不能凭空产生，也不能凭空消失。

NONE OF THESE DEVOURINGS CAN CREATE ANY NEW ENERGY: THEY JUST PASS ALONG ENERGY THAT CAME FROM SOMEWHERE ELSE. AND WHERE IS THAT? IF YOU TRACE BACK THE LINKS OF ANY FOOD CHAIN FAR ENOUGH, YOU'LL COME TO A ***PLANT,*** WHICH GOT ITS ENERGY STRAIGHT FROM THE ***SUN.*** ALL LIFE DEPENDS ULTIMATELY ON THE SUN.*

狼吞虎咽地吃都不能产生新的能量，而是将其他地方的能量相互传递。其他地方是哪儿呢？如果追溯任何一个食物链的各个环节，其源头一定是某种**植物**，这种植物直接从**太阳**那里得到能量。所有的生命体最终都要依靠太阳。*

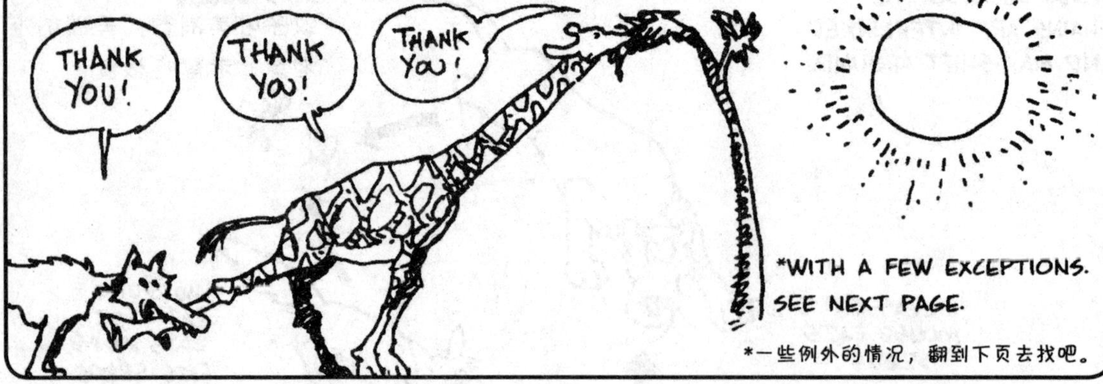

*WITH A FEW EXCEPTIONS. SEE NEXT PAGE.

*一些例外的情况，翻到下页去找吧。

UNTIL HUMANITY CAME ALONG, THERE WERE ONLY TWO WAYS FOR LIVING BEINGS TO GET ENERGY: SIT IN A WARM PLACE, OR EAT SOMETHING!

在人类出现之前，生物体获得能量的方式只有两种：待在温暖的地方或者吃些东西。

OF THE TOTAL SUNLIGHT FALLING ON EARTH, ***30%*** IS REFLECTED, NEARLY ***50%*** IS CONVERTED TO HEAT, AND ALMOST ALL THE REST POWERS THE WATER CYCLE—EVAPORATION, RAIN, WIND, ETC. ***LESS THAN 1%*** IS USED BY LIVING PLANTS.

照射到地球表面全部太阳光的30%被反射，近50%转化成热量，剩下的给水循环提供动力——蒸发、降雨、使空气流动等。生长的植物只用了**不到1%**。

BUT THAT TINY FRACTION OF SOLAR ENERGY PROVIDES ALL OF LIFE'S FOOD NEEDS, THROUGH THE PROCESS OF

PHOTOSYNTHESIS.

但是，太阳能那微小的一部分通过**光合作用**的过程，为所有生命提供了所需要的食物。

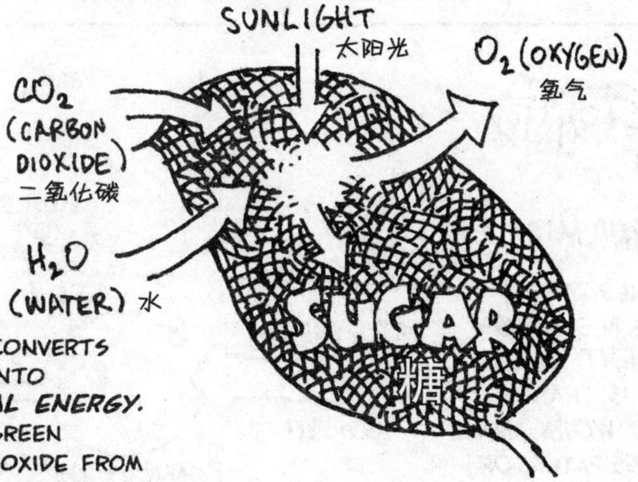

PHOTOSYNTHESIS CONVERTS ***SOLAR ENERGY*** INTO ***STORED CHEMICAL ENERGY.*** IN THE CELLS OF GREEN PLANTS, CARBON DIOXIDE FROM THE AIR, WATER FROM THE EARTH, AND LIGHT FROM THE SUN REACT TO PRODUCE ***SUGAR,*** A COMPLEX ORGANIC COMPOUND THAT STORES CHEMICAL ENERGY FOR LATER USE. OXYGEN IS RELEASED AS A BY-PRODUCT.

光合作用将**太阳能**转化为**储存的化学能**。在绿色植物的细胞中，空气中的二氧化碳、来自土壤的水和太阳光之间发生反应产生了糖。糖是一种可以储存化学能的复杂有机物。氧气作为副产物被释放出来。

THIS STORED CHEMICAL ENERGY DRIVES ALL THE BIOGEOCHEMICAL CYCLES OF THE EARTH.

储存的化学能推动地球上所有的生物地球化学循环。

EXCEPT: AT ***UNDERSEA VOLCANIC VENTS,*** SULFUR-LOVING BACTERIA CAN CONVERT THE ***EARTH'S HEAT*** INTO CHEMICAL ENERGY, A PROCESS CALLED ***CHEMOSYNTHESIS.*** THESE BACTERIA SUPPORT A COMMUNITY OF WORMS, CRABS, AND CLAMS IN THE TOTAL DARKNESS.

特例：在**海底火山口**，嗜硫细菌能将**地球热量**转化为化学能。这个过程叫作**化能合成**。这些细菌在完全黑暗的环境中可以供养蠕虫、蟹类和蛤蚌群落。

VOLCANIC HEAT COMES FROM THE ENERGY OF ***RADIOACTIVE DECAY*** OF ELEMENTS IN THE EARTH.

火山的热量来自地球上一些元素的放射性衰变后产生的能量。

THE SECOND LAW OF THERMODYNAMICS

SAYS THAT ENERGY CONVERSIONS ARE ***NEVER 100% EFFICIENT:*** WHENEVER ENERGY IS TRANSFORMED INTO ***WORK,*** SOME IS ALWAYS DISSIPATED, OR WASTED, AS HEAT.

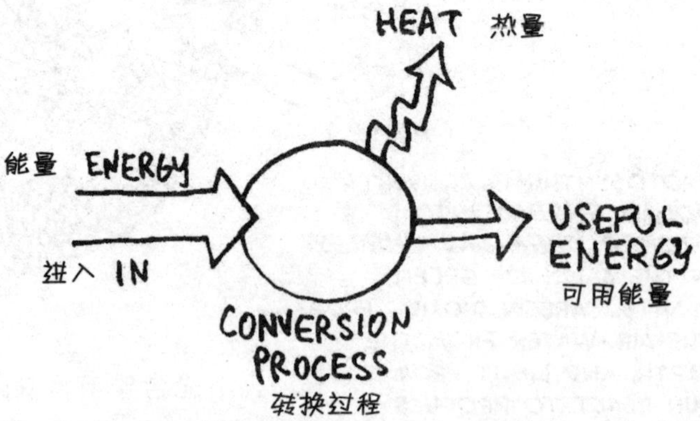

热力学第二定律指出，能量不能100%**被转化**：任何时候能量转化为**功**，总会有一部分能量以热量形式耗散或浪费。

FOR EXAMPLE, WHEN A CAR CONVERTS THE CHEMICAL ENERGY OF GASOLINE INTO MOTION, MOST OF THE ENERGY TURNS TO HEAT: THE ENGINE AND EXHAUST GASES GET HOT, FRICTION HEATS THE WHEEL BEARINGS, ETC... ONLY ABOUT *15%* OF THE ORIGINAL CHEMICAL ENERGY ACTUALLY MOVES THE CAR!

例如，当一辆汽车将汽油产生的化学能转化为动能时，大多数能量变成热能：发动机和废气变热，摩擦使车轮轴承变热，等等。实际上，只有大约15%的原化学能用来驱动汽车！

ENERGY IN = WORK + HEAT
AND OTHER WASTED ENERGY

能量 = 功 + 热能
以及其他损失的能量

ECOLOGICAL EFFICIENCY,

OR ***FOOD CHAIN EFFICIENCY,*** IS THE PERCENTAGE OF USABLE ENERGY CAPTURED AT EACH LEVEL OF CONSUMPTION. FOR EXAMPLE, PLANTS RANGE IN EFFICIENCY BETWEEN ***1%*** AND ***3%*** DEPENDING ON THE PLANT: ONLY 1–3% OF THE SOLAR ENERGY ABSORBED BY THE PLANT IS ACTUALLY CONVERTED TO BIOMASS.

生态效率或者**食物链效率**是指在每级消费中捕获可用能量的百分比。例如，植物的效率在1%～3%之间，因植物物种的不同而不同：实际上植物吸收的太阳能中只有1%～3%被转化为生物量。

A TYPICAL HERBIVORE USES SOME ***10%*** OF THE TOTAL PLANT ENERGY CONSUMED, WITH THE REST LOST TO HEAT OR RESPIRATION.

典型的食草动物所吃的植物作为能量只有10%被利用，而剩下的以热量和呼吸的方式损失掉。

我需要吃很多草……

A CARNIVORE'S EFFICIENCY IS ALSO ABOUT ***10%,*** MEANING THE CARNIVORE GETS ONLY 1/10 OF 1/10 OF THE ORIGINAL PLANT ENERGY.

食肉动物的效率也是大约10%，这就意味着肉食动物仅仅利用了最初1/10的植物能量中的1/10。

而我需要的草是你的10倍！

THE TOTAL EFFICIENCY AT ANY LEVEL OF CONSUMPTION IS THE PRODUCT OF THE EFFICIENCIES OF ALL THE CONVERTORS UP TO THAT LEVEL. IN THIS CASE (TAKING THE PLANT EFFICIENCY AS 2%), THE CARNIVORE'S TOTAL EFFICIENCY IS

任何一级食物消耗的总效率是各级消耗效率的乘积。那么（假设植物效率是2%），食肉动物的总效率就是

$$0.02 \times 0.1 \times 0.1 = 0.0002.$$

THE CARNIVORE USES ONLY ***0.02%*** OF THE SOLAR ENERGY THAT WENT INTO THE GRASS THAT WENT INTO THE COW THAT WENT INTO THE CARNIVORE!

食肉动物仅仅利用了太阳能的0.02%，太阳能被草吸收，草的能量被牛吸收，牛肉的能量被食肉动物吸收！

WE CAN ALSO THINK OF THE LINKS IN THE FOOD CHAIN IN TERMS OF LEVELS, WITH EACH LEVEL EATING FROM THE ONE BELOW. THE FIRST **TROPHIC* LEVEL** CONSISTS OF GREEN PLANTS, AS WELL AS PHOTO-SYNTHESIZING AND CHEMO-SYNTHESIZING BACTERIA.

食物链的链环也可以用层级的概念来思考，每个层级的食物都以各自的下一层级为食。第一**营养级***包含绿色植物、光合细菌和化能合成细菌。

*FROM GREEK *TROPHIKOS* = NOURISHMENT, AS IN ENGLISH "TROUGH."

*来自希腊语"Trophikos"即营养，就像英语单词"trough"。

THE SECOND TROPHIC LEVEL CONSISTS OF HERBIVORES OR PLANT-EATERS. THEY ARE ALSO CALLED ***PRIMARY CONSUMERS.***

第二营养级为食草动物或者说植物消费者。这些动物也叫初级消费者。

PRODUCER SPECIES ARE ALSO CALLED ***AUTOTROPHS***, SINCE THEY MAKE THEIR OWN FOOD FROM CARBON DIOXIDE AND WATER.

生产者物种也叫作**自养生物**，因为它们从二氧化碳和水中制造自己的食物。

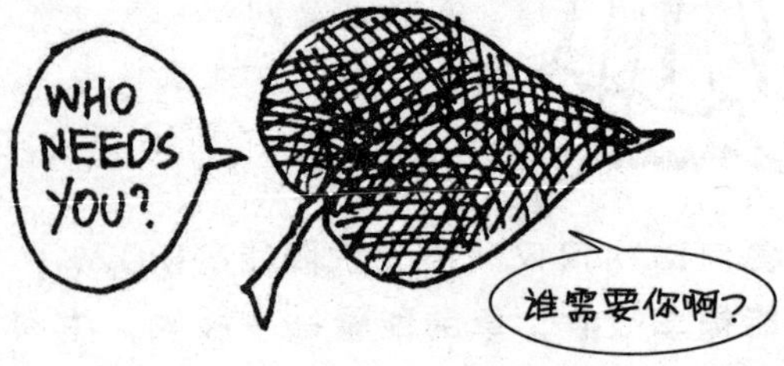

NONPLANT ***HETEROTROPHS*** CAN'T MAKE THEIR OWN FOOD, SO THEY ARE CONSUMERS OF PLANTS OR OTHER HETEROTROPHS.

非植物**异养生物**不能制造自己的食物。因此，它们是植物或者其他异养生物的消费者。

THE THIRD TROPHIC LEVEL CONSISTS OF CARNIVORES THAT EAT HERBIVORES— ALSO CALLED ***SECONDARY CONSUMERS***... AND IN SOME ECOSYSTEMS THERE MAY BE LEVELS OF CARNIVORES EATING CARNIVORES: ***TERTIARY*** OR ***QUATERNARY*** CONSUMERS.

第三营养级是那些以食草动物为生的食肉动物，也叫**次级消费者**……在一些生态系统中，也可能存在食肉动物食用其他食肉动物的层级，即**三级**消费者或**四级**消费者。

THERE ARE ALSO ***OMNIVORES*** THAT EAT FROM MORE THAN ONE TROPHIC LEVEL AT A TIME.

同时还有**杂食动物**，他们会同时食用多个营养层的食物。

THESE STACKED LAYERS OF EATERS ARE ***MACROCONSUMERS***, I.E., THEY'RE BIG.

这些不同层级的食用者属于**大型消费者**，也就是说，他们都是大家伙。

THERE ARE ALSO ***MICRO-CONSUMERS***, OR DECOMPOSERS, MOSTLY BACTERIA OR FUNGI.

同时也存在**微型消费者**，或者叫分解者，它们大多数是细菌或真菌类。

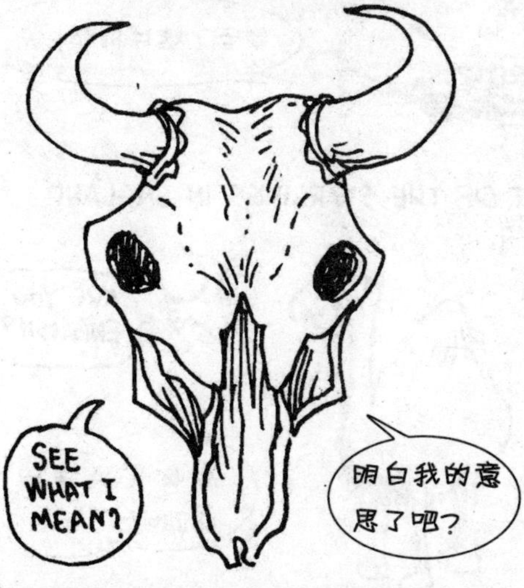

IN MANY TERRESTRIAL ECOSYSTEMS, AS MUCH AS 90% OF ALL PLANT MATTER FALLS STRAIGHT INTO THE DECOMPOSER FOOD WEB.

在许多陆地生态系统中，高达90%的植物直接落入分解者的食物网中。

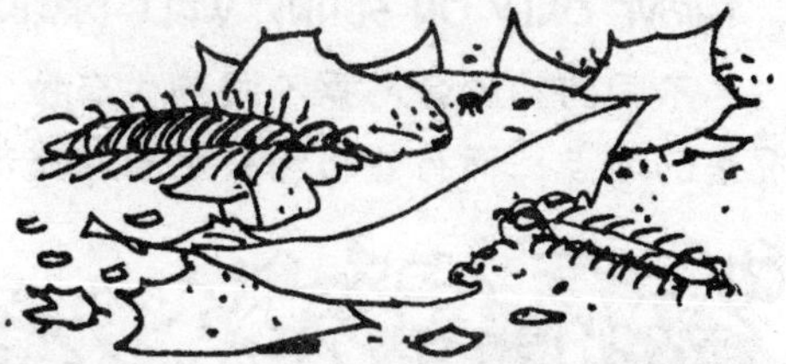

EARLY STAGES MAY INVOLVE MILLIPEDES, WEEVILS, ETC., BUT THE FINAL BREAKDOWN IS DONE BY BACTERIA AND FUNGI.

早期的分解者可能有千足虫、象鼻虫等。但最终的分解工作是由细菌和真菌完成的。

EACH SPECIES HAS TO CARVE OUT ITS PLACE IN AN ECOSYSTEM. THIS MEANS FINDING A HABITAT, TOLERABLE LEVELS OF WATER, LIGHT, AND TEMPERATURE, AND A SUPPLY OF AVAILABLE FOOD. THIS FAVORABLE COMBINATION OF FACTORS IS CALLED A

NICHE.

每个物种必须在生态系统中划出自己的地盘。这就意味着要找到一个栖息地，拥有好的水源、光线、温度和足够的食物。这些有利条件的结合被称为一个**生态位**。

DIFFERENT SPECIES TEND TO OCCUPY DIFFERENT NICHES. SOME PLANTS, FOR EXAMPLE, PREFER SHADY, MOIST, POORLY DRAINED SPOTS, WHILE OTHERS THRIVE ONLY ON SUNNY, WELL-DRAINED HILLSIDES.

不同的物种会占据不同的生态位。例如，有些植物更喜欢成荫的、潮湿的、排水状况差的地点；而有些植物只生长在充满阳光、排水状况好的山腰地带。

A NICHE CAN BE HIGHLY SPECIALIZED, LIKE THAT OF THE STARLINGS IN ENGLAND THAT EAT ONLY THE TICKS OFF SHEEP AND DEER.

有的生态位可能会非常特别。比如说英国的椋鸟只吃绵羊和鹿身上的蜱虫。

WHEN RESOURCES ARE ABUNDANT, NICHES CAN BE SO BROAD THEY OVERLAP, BUT WHEN RESOURCES ARE LIMITED (THE USUAL SITUATION), OVERLAPPING NICHES MEANS **COMPETITION.**

资源丰富时，生态位可能会很广阔，不同物种的生态位可能会重叠；而资源有限时（资源往往是有限的），生态位的重叠也就意味着**竞争**。

WHEN TWO SPECIES COMPETE FOR THE SAME SCARCE RESOURCES, ONE SPECIES GENERALLY WILL ***TAKE OVER THE NICHE*** AND EXCLUDE THE OTHER, WHICH HAS TO FIND SOME OTHER WAY TO FEND FOR ITSELF. THIS IS THE **PRINCIPLE OF COMPETITIVE EXCLUSION.**

当两个物种争夺稀少的资源时，通常其中一个物种会**占据生态位**并排挤掉另一个物种，使其不得不去寻找其他养活自己的办法。这被称为**竞争排斥原理**。

IF RESOURCES PERMIT, TWO SPECIES MAY SPECIALIZE AND ***PARTITION***, OR SPLIT, A NICHE INTO TWO SMALLER ONES. THE SHAG AND THE CORMORANT ARE BOTH FISH-EATING DIVING BIRDS, BUT THE CORMORANT DIVES DEEPER, LEAVING THE FISH ON TOP TO THE SHAG.

假如资源允许，两个物种可能会变得特化，并将一个生态位**分割**为几个小的生态位。欧鸬鹚和普通鸬鹚都是能潜水的食鱼鸟，但是普通鸬鹚潜得更深，将海水表层的鱼留给了欧鸬鹚。

WHEN TWO SPECIES REACH SUCH AN ACCOMMODATION, THEY MAY EVOLVE ***SPECIALIZED FEATURES.*** FOR INSTANCE, IF TWO SPECIES OF SIMILAR BIRDS COEXIST, ONE MAY EVOLVE A LONG THIN BILL TO EAT LARGE BUGS, WHILE THE OTHER WILL HAVE A TOUGH, STUBBY BILL GOOD FOR CRACKING SEEDS. THIS IS CALLED ***CHARACTER DISPLACEMENT.***

当两个物种达成这样的和解时，它们就可能进化出**独有的特征**。例如，如果两种相似的鸟共存，其中一种能进化出细长的喙以便进食大虫，而另一种则会有一个短且粗硬的喙以便啄裂种子，这被称为**性状替换**。

SOMETIMES, NO ACCOMMODATION IS POSSIBLE. ONE SPECIES HAS A ***COMPETITIVE ADVANTAGE:*** IT'S FASTER, STRONGER, OR MORE TOLERANT OF CHANGES IN SOME LIMITING FACTORS... WITH THE RESULT THAT THE LESS WELL ADAPTED SPECIES IS DRIVEN TO COMPLETE ***EXTINCTION.***

有时当没有和解可言时，一个物种具有的**竞争优势**：如跑得更快、力气更大，对适应环境因子的变化更迅速……那么，那些适应能力弱的物种就彻底**灭绝**了。

SOME LESS COMPETITIVE WAYS FOR SPECIES TO INTERACT INCLUDE:

有些竞争手段不那么激烈，但可以供物种相互影响。例如：

SYMBIOSIS

("LIVING TOGETHER") IS A LONG-TERM RELATIONSHIP IN WHICH TWO SPECIES EXCHANGE ENERGY OR ADAPTIVE BENEFITS. THERE ARE THREE KINDS OF SYMBIOSIS, DEPENDING ON WHICH SPECIES RECEIVES THE MOST BENEFIT.

共生（也就是“一起生活”）是两个物种为交换能量或者互相利用各自优点所建立的长期共存关系。根据物种得到的利益，共生关系可分为三类。

IN ***COMMENSALISM,*** ONE SPECIES GAINS, WHILE THE OTHER IS ESSENTIALLY UNAFFECTED. FOR EXAMPLE, WE EAT NUTS, BUT THE NUT TREE BARELY NOTICES.

第一类是**共栖**。在**共栖**的过程中，一个物种获利，而另一个物种实际上不受影响。例如，我们从树上摘核桃吃，而核桃树从不抗议。

IN ***MUTUALISM,*** BOTH SPECIES BENEFIT. WE SAW AN EXAMPLE OF MUTUALISM IN ***CORAL,*** WHICH PROVIDES SHELTER TO MICRO-ORGANISMS THAT SECRETE NUTRIENTS USEFUL TO THE CORAL.

第二类是**互惠共生**。在此关系中，两个物种同时受益。以**珊瑚**为例，珊瑚为微生物提供栖息之处，同时这些微生物为珊瑚提供营养物。

IN ***PARASITISM,*** THE PARASITE HARMS ITS HOST, SUCKING AWAY ITS ENERGY, USUALLY SLOWLY. EXAMPLES ABOUND: LEECHES, FLEAS, TICKS, INFECTIONS BY BACTERIA, PROTOZOA, FUNGI, OR WORMS, ETC..

第三类是**寄生关系**。在**寄生关系**中，寄生物伤害宿主，耗尽它的能量，这通常是个很缓慢的过程。这类例子很多，如蛭类、跳蚤、蜱虫以及细菌感染、真菌感染、原虫或者蠕虫感染等。

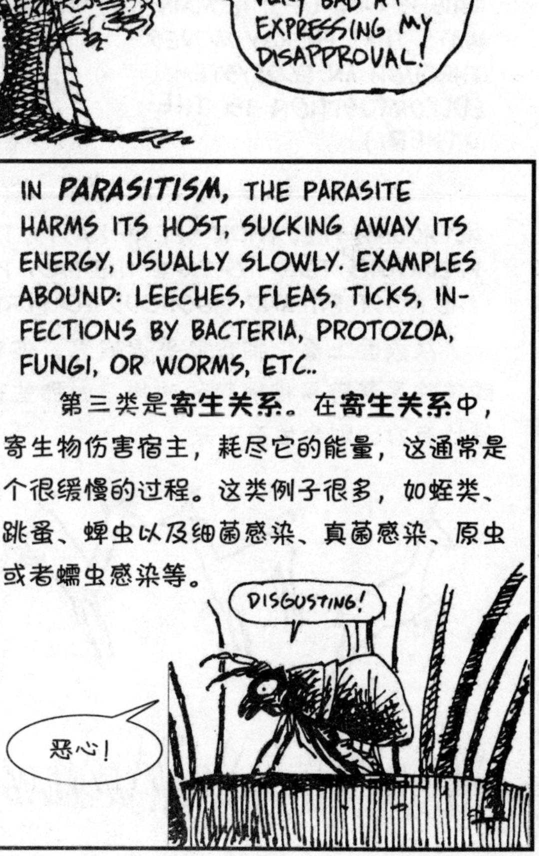

PREDATION

IS A RATHER SIMPLE-MINDED RELATIONSHIP: ONE SPECIES, THE PREDATOR, ***HUNTS AND EATS*** THE OTHER SPECIES, THE PREY.

捕食关系是一种相当简单的关系：一个物种作为捕食者**捕捉并吃掉**另一个物种，也就是猎物。

THIS IS ONE OF THE MAIN WAYS THAT ENERGY MOVES THROUGH AN ECOSYSTEM. (DECOMPOSITION IS THE OTHER.)

这是能量在生态系统中转移的主要方式之一（另一种方式是分解）。

ALTHOUGH PREDATION SEEMS NASTY, IT DOES BENEFIT THE PREY SPECIES. PREDATORS TEND TO TAKE THE EASY PICKINGS, THE SLOW AND INFIRM, LEAVING THE MOST FIT AND VIGOROUS TO SURVIVE.

从表面上看，捕食似乎很残忍，但它对猎物物种是有益的。因为捕食者吃掉的往往是最容易被捕获的动物，如那些行动缓慢的、骨瘦如柴的动物等。而那些健壮有力的则会生存下来。

PREDATOR AND PREY POPULATIONS TEND TO BE IN ***DYNAMIC BALANCE.*** OVERPREDATION REDUCES PREY POPULATIONS, SO PREDATORS DIE OF STARVATION, ALLOWING PREY POPULATIONS TO REBOUND. THEN PREDATORS FLOURISH, OVEREAT AGAIN, AND DIE OFF, ETC. A FAMOUS GRAPH BY LOTKA-VOLTERRA SHOWS HOW THESE TWO POPULATIONS RISE AND FALL WITH EACH OTHER, SEPARATED BY A TIME LAG.

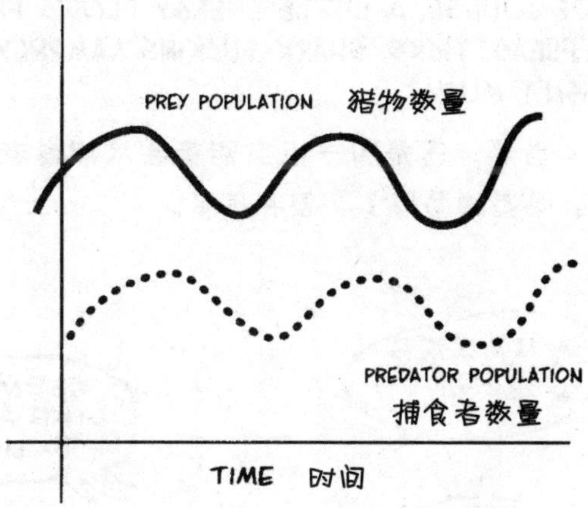

捕食者及其猎物的数量往往呈现**动态平衡**。过度捕猎减少了猎物的数量，最终捕食者也被饿死，这就使猎物的数量又重新反弹，为捕食者提供食物。这样捕食者重新兴旺起来，再次过度捕猎，再度饿死，如此轮回反复。洛特卡-沃尔泰拉的著名曲线图展示了捕食者和猎物物种的数量随时间变化而变化的规律。

THE PREDATOR SPECIES PROWL THE TOP OF THE FOOD CHAIN. THE FLOW OF FOOD ENERGY MOVES UPWARD FROM PLANT TO PREDATOR, AND THERE IT STOPS. (ON LAND, PREDATORS RARELY PREY ON OTHER PREDATORS; UNDERWATER, THERE MAY BE MORE TROPHIC LEVELS, BUT THEY HAVE TO END SOMEWHERE.)

捕食者物种总是在食物链的最高层。食物能量从植物向上流动到捕食者，然后停止（在陆地上，捕食者很少捕猎其他捕食者；而水下可能会有更多的营养级，但最终也仍然会止于某级）。

OF COURSE, A LITTLE ENERGY FLOWS FROM PREDATORS TO THEIR PARASITES (FLEAS, TICKS, SHARK-SUCKING LAMPREYS, INTESTINAL BACTERIA, ETC.) BUT IT ISN'T MUCH.

当然，还是有一点点能量会从捕食者流向其寄生物（跳蚤、蜱虫、食鲨七鳃鳗、肠道细菌等），但并不多。

ONLY AT DEATH DOES THE PREDATOR'S CHEMICAL ENERGY RETURN TO THE ECOSYSTEM VIA THE DECOMPOSERS AND SCAVENGERS.

捕食者只有在死亡后，其化学能才会通过分解和食腐动物返回到生态系统中去。

AMONG THE LIMITING FACTORS FOR PREDATORS ARE THESE:

维持肉食动物生存的限制因子有：

*裸鼹鼠：群居在非洲等地，皮肤上几乎无毛，没有疼痛感，长得极其丑陋，但因体内含癌症病毒的天然抗性物质而深受科学家喜爱。文中狮子因其长像丑陋而不屑于吃它。

NOW TRY TO IMAGINE A ***SUPERPREDATOR*** THAT COULD SOMEHOW ***OVERCOME*** THOSE LIMITING FACTORS... A SPECIES THAT CONTINUALLY ***IMPROVES ITS HUNTING TECHNIQUE*** AND IS WILLING TO EAT NEARLY ***ANYTHING...*** AND WHAT IF THIS SPECIES ALSO DISCOVERED ***SECRET SOURCES OF ENERGY*** THAT HAD ALWAYS BEEN HIDDEN FROM ALL OTHER ANIMALS? TRY TO IMAGINE IT...

试着想象一下，有那么一只**超级捕食者**，它可以**克服**所有这些限制因子，能够不断**提高捕猎技巧**，并且愿意吃任何东西……而且，这个物种还发现了**秘密的能量源**，这个能量源从来不为其他动物所知，那么，世界会是什么样子呢？试着去想象一下吧……

· CHAPTER 7 ·

FROM HUNTING TO PLANTING

·第七章·

从狩猎到种植

LIKE ALL CREATURES, WE ARE PART OF THE WEB OF LIFE. WE EAT FOOD FROM EVERY ***TROPHIC LEVEL***; WE HAVE FEW ***PREDATORS***; AND WE HAVE MANY ***SYMBIOTIC*** RELATIONSHIPS: WE PRACTICE ***MUTUALISM*** WITH THE BACTERIUM *E. COLI*, WHICH HELPS US DIGEST FOOD IN EXCHANGE FOR A SAFE, WARM HOME IN OUR INTESTINES; WE HAVE ***PARASITIC*** DISEASES; AND WE HAVE ***COMMENSAL*** RELATIONSHIPS WITH PIGEONS AND COCKROACHES (THEY THRIVE; WE'RE UNAFFECTED).

和所有的生物一样，我们也是生命网中的一部分。我们从每个**营养级**获取食物；很少被捕食者捕食；我们有很多**共生**关系：如与大肠杆菌有互惠共生关系，它们帮助我们消化食物，交换条件是在我们的肠子里有一个温暖而安全的栖身地；我们会患上**寄生虫引起的**疾病；我们与鸽子、蟑螂有**共栖**的关系（它们繁殖生存，而我们不受影响）。

OF COURSE, OUR WAY OF LIFE HAS CHANGED OVER THE YEARS. IN THE BEGINNING, SOME 2 MILLION YEARS AGO, OUR ANCESTORS OCCUPIED THE NICHE OF ***HUNTER-GATHERERS,*** WHICH REMAINED HUMANITY'S NICHE FOR 99% OF OUR HISTORY.

当然，随着时间的推移，我们的生存方式也改变了。大约200万年前，我们的祖先最早靠狩猎和采集为生，占据着**狩猎—采集者**的生态位。在人类历史长河99%的历程中，都始终保持着这一生态位。

STUDIES OF MODERN SOCIETIES SUGGEST THAT HUNTING AND GATHERING IS SURPRISINGLY EASY. FOOD IS ABUNDANT IF YOU KNOW WHAT TO LOOK FOR, SO HUNTER-GATHERERS DON'T HAVE TO WORK VERY HARD. THEY HAVE PLENTY OF TIME FOR PLAYING, PERFORMING RITUALS, OR DISCUSSING FAMILY ISSUES.

对现代社会的研究可见，狩猎和采集极其简单。如果知道想找什么的话，食物会很丰富，因此狩猎—采集者不必太勤劳。他们拥有很多时间来玩、参加礼仪社交活动或者讨论家务事。

HUNTER-GATHERER POPULATIONS ARE SMALL, RARELY EXCEEDING 50... THEY HAVE TO BE MOBILE, SO POSSESSIONS ARE A BURDEN... AND POPULATION CONTROL IS PRACTICED IN THE FORMS OF INFANTICIDE, ABANDONMENT OF THE AGED AND INFIRM, AND EVEN MURDER.

狩猎—采集者的种群很小，基本不超过50人……他们总要不断迁移，所以随身携带的物品就成了负担……人口增加往往也成为负担，因此他们通过杀婴、抛弃年老体弱者甚至杀戮来控制人数。

IF A GROUP GROWS TOO LARGE FOR ITS TERRITORY TO SUSTAIN IT, A FACTION SPLITS OFF AND MOVES AWAY.

如果一个群落的人口增长过多，该土地上的人口就难以维持生计，一部分人就要脱离这个群落，远走他乡。

WHAT CHANGED THIS STABLE IF NOT ENTIRELY IDYLLIC PICTURE? SURELY IT WAS THE MASTERY OF

FIRE.

这种生活方式虽非田园生活般惬意，但至少很稳定。那又是什么改变了这种生活方式呢？可以肯定地说是火的使用。

FIRE WAS THE FIRST OF THE ***HIDDEN ENERGY SOURCES*** LIBERATED BY HUMANS. WHAT ANIMAL (ASIDE FROM TERMITES) EVER THOUGHT OF GETTING USEFUL ENERGY FROM ***WOOD?*** WITH THIS EXTRA, PORTABLE, NONFOOD ENERGY, PEOPLE COULD SUDDENLY LIVE ANYWHERE. CLIMATE WAS NO BARRIER...

火是人类开发的第一种**隐藏能源**。还有什么动物（除了白蚁外）曾想过从树木里得到有用的能量呢？有了这种额外的、可携带的而又不是食物的能量，人们突然能够在任何地方居住，气候再也不是障碍了……

(ESPECIALLY AFTER THEY INVENTED ***CLOTHES***, THE FIRST ENERGY-CONSERVATION TECHNOLOGY*).

（特别是在他们发明了**衣服**后，即发明了第一个能量保存技术*之后，更不用受气候限制了。）

*OR MAYBE THE SECOND. HOUSES ALSO RETAIN HEAT.

*衣服也可能是第二个能量保存技术，因为房屋也能保存热量。

SO BEGAN THE FIRST HUMAN ***POPULATION EXPLOSION***, AS OUR ANCESTORS, NOW MORE EVOLVED, SPREAD OVER THE WHOLE WORLD.

我们的祖先进一步进化，人类也遍及世界，于是，第一次**人口大爆炸**开始了。

BY THE END OF THE LAST ICE AGE, HUMANS HAD EVOLVED TO THE POINT THAT THEY COULD DO SOME REAL DAMAGE... EQUIPPED WITH ADVANCED TECHNOLOGY (BOWS & ARROWS, SNARES, BASKETS, AN ELABORATE STONE TOOLKIT), HOMO SAPIENS BECAME A ***SUPERPREDATOR***. THE HUNTERS USED FIRE TO FLUSH GAME, DRIVING WHOLE HERDS OVER CLIFFS.

到了最后的一个冰期的末期，人类进化成为真正的破坏者。由于拥有先进的技术（弓箭、陷阱、笼子、精致的石器工具包等），人类成为**超级捕食者**。狩猎者用火惊吓并驱赶整个畜群，把它们逼下悬崖。

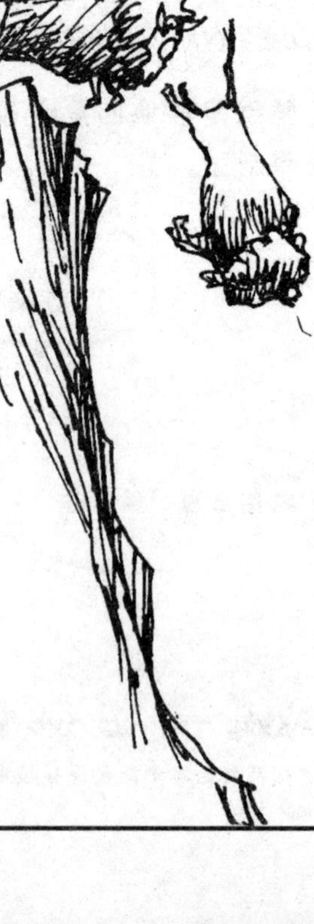

BY 10,000 YEARS AGO, OVER ***70%*** OF THE LARGE LAND ANIMALS WERE EXTINCT—BY HUMAN HANDS, ACCORDING TO THE THEORY OF ***PLEISTOCENE OVERKILL.***

一万年前，超过70%的大型陆生动物都灭绝了。根据**更新世*大屠杀理论**，这都出自人类之手。

*更新世：地质年代名称。

NOT EVERYONE ACCEPTS THE THEORY OF PLEISTOCENE OVERKILL. THESE EXTINCTIONS ALSO COINCIDED WITH THE END OF THE LAST ICE AGE, AND MANY SCIENTISTS CONTEND THAT THE ANIMALS DIED AS A RESULT OF CLIMATIC CHANGE. BUT IT IS HARD TO IMAGINE THAT PEOPLE PLAYED NO ROLE IN TIPPING THE BALANCE AGAINST ALL THOSE MAMMOTHS, GROUND SLOTHS, GIANT ELK, ETC.

并非所有人都赞成更新世大屠杀理论。动物的大批灭绝正好处于最后一个冰期的末期。许多科学家争辩说，那些动物是因为气候变化而亡的，然而，猛犸象、地獭、巨鹿的数量失衡，很难想象不是人类在搅局。

THIS MUST HAVE BEEN A SERIOUS CRISIS FOR HUMANITY: WITH A POPULATION IN THE NEIGHBORHOOD OF 4 MILLION, OUR SPECIES WAS SUDDENLY FACED WITH A SHORTAGE OF BIG GAME...

这对人类来说势必已经成为一个严重危机：由于大猎物数量急剧减少，400万人口突然面临生计问题。

BUT OUR ANCESTORS WERE INTELLIGENT, SOCIAL OMNIVORES WITH SOME TECHNOLOGY, AND THEY WERE ABLE TO TURN A CRISIS INTO AN OPPORTUNITY.

但是我们的祖先很聪明，是社会杂食性动物。他们身怀“绝技”，甚至有化危机为机遇的本领。

THEY INVENTED **AGRICULTURE**, USHERING IN THE MOST FUNDAMENTAL UPHEAVAL IN HUMAN HISTORY, AND POSSIBLY IN ALL WORLD HISTORY.

我们的祖先创造了**农业**，开始了人类历史最根本的变革，并很可能开启了整个世界历史的新篇章。

WHAT IS AGRICULTURE? IN TERMS OF THE FOOD WEB, FARMING MEANS A RADICAL ***SIMPLIFICATION*** AND ***REORGANIZATION*** OF THE FLOW OF ENERGY. THE FARMER UPROOTS (OR CHASES AWAY, OR BURNS OFF) UNWANTED SPECIES AND SOWS ONLY THOSE CROPS THAT ARE FIT FOR HUMAN CONSUMPTION.

什么是农业？用食物网的概念，耕作意味着能量流动的**高度简化和重组**。这意味着拔除不需要的物种（也可以驱除或者焚烧），只播种人类消费所需的农作物。

IN THIS WAY, ALL (OR NEARLY ALL) THE PLANT ENERGY ON A GIVEN PLOT OF GROUND NOW FLOWS INTO HUMAN MOUTHS AT THE EXPENSE OF THE OTHER SPECIES THAT USED TO FIND A NICHE THERE.

这样，一块土地上几乎所有的植物能量现在都流入了人类口中，而牺牲了原本在这里生存的其他物种。

WITH THE INVENTION OF FARMING, HUMAN SOCIETY EXPERIENCED A SUDDEN INFLUX OF ENERGY. THIS HAD TWO MAJOR EFFECTS: ***POPULATION GROWTH*** AND ***SOCIAL ORGANIZATION.***

随着农业的发明，人类社会经历了能量的突然涌入。这产生了两个主要影响：**人口增长**和**社会组织出现**。

POPULATION GREW BECAUSE AGRICULTURE INCREASES THE CARRYING CAPACITY OF THE LAND: THERE WAS MORE FOOD FOR HUMANS, HENCE MORE HUMANS!

由于农业生产提高了土地的使用率，人口也随之增长。也就是说，供人类食用的粮食增加了，人口也就增加了！

UNLIKE THE HUNTER-GATHERER, THE FARMER WANTS A LARGE FAMILY. MORE CHILDREN MEAN MORE FIELD HANDS, HENCE MORE FOOD, NOT LESS! (THIS MENTALITY IS ONE OF THE OBSTACLES TO POPULATION CONTROL IN MAINLY RURAL COUNTRIES TODAY.)

与狩猎者和采集者不同的是，农民喜欢大家庭，因为孩子越多，干农活的人就越多，生产的粮食也就越多！（这种心态是当今农业国家控制人口的障碍之一）

IN GENERAL, WHENEVER ANY COMPLEX SYSTEM HAS A SUPPLY OF ENERGY COMING IN, THE SYSTEM MAY TEND TO **ORGANIZE ITSELF.** (THIS BEAUTIFUL IDEA IS DUE TO THE NOBEL PRIZE-WINNING CHEMIST ILYA PRIGOGINE.)

一般来说，每当一个复杂系统有能量供应时，这个系统往往会自我组织。（这个伟大的理论来自诺贝尔化学奖获得者伊利亚·普里高津。）

THE INFLUX OF FOOD ENERGY FROM AGRICULTURE HAD A SIMILAR EFFECT WHEREVER IT HAPPENED: DIVISION OF LABOR... SOCIAL HIERARCHIES... GOVERNMENT COMMITTEES... TEMPLES... MARKETS...

农业生产带来的粮食供应也会产生同样的效应：如劳动力的分工、社会等级、政府机构、寺庙、市场的产生等。

AND ULTIMATELY, IN SOME CASES,

CIVILIZATION.

最终，在某些情形下，**文明**出现了。

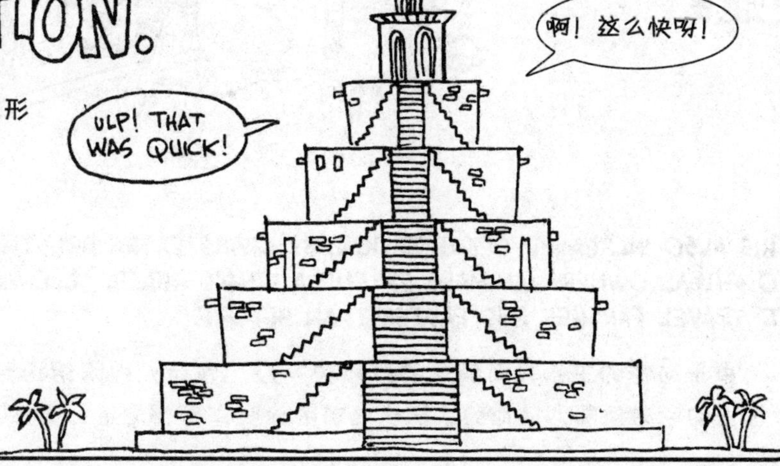

IN MOST PLACES WHERE AGRICULTURE DEVELOPED, PEOPLE ALSO DOMESTICATED ANIMALS. THIS HAD A SIMILAR EFFECT: A HERDER'S FLOCK USES MOST OF THE AVAILABLE PLANT ENERGY, LEAVING LESS FOR WILD ANIMALS, INCLUDING PREDATORS.

在大多数农业发展的地方，人们也会饲养动物。这有一个相似的影响：放牧人的牲畜吃掉大部分可获得的植物能量，很少一部分留给了野生动物，包括捕食者。

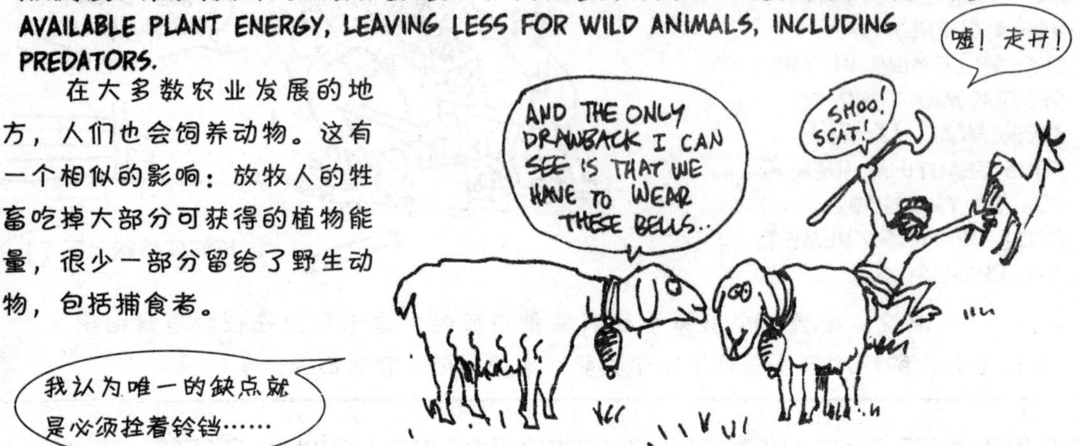

DOMESTICATED PLANTS AND ANIMALS BOTH LIVE IN SYMBIOTIC RELATIONSHIPS WITH PEOPLE: WE PROTECT THEM, THEY FEED US.

家养植物和动物都和人类有共生关系：我们养着它们，它们喂饱我们。

BESIDES EATING ANIMALS, PEOPLE ALSO HARNESSED THEM, ADDING ANIMAL POWER TO OUR ENERGY RESOURCES.

除了把动物当作食物外，人类还役使动物，把动物的力气当作能量来源。

THIS ALSO INCREASED ORGANIZATION, BY GIVING EXTRA PRESTIGE AND POWER TO ANIMAL OWNERS. ANIMALS EXTEND A KING'S REACH, ALLOWING HIS ARMIES TO TRAVEL FARTHER AND FASTER THAN BEFORE.

由于动物的主人具有额外的声望和权力，因而，组织结构更加坚固。动物能够扩大国王的活动范围，使他的军队比之前的赤脚军行得更远、更快。

THE FIRST FARMERS LIVED IN THE HILLS OF WESTERN ASIA AROUND 8000 B.C., WHERE THEY DOMESTICATED SHEEP AND GOATS AND BRED BETTER STRAINS OF WHEAT AND BARLEY.

大约公元前8000年，第一批农民生活在西亚的丘陵地区，在那里，他们同时也驯养山羊和绵羊，并培育出更好的小麦和大麦品种。

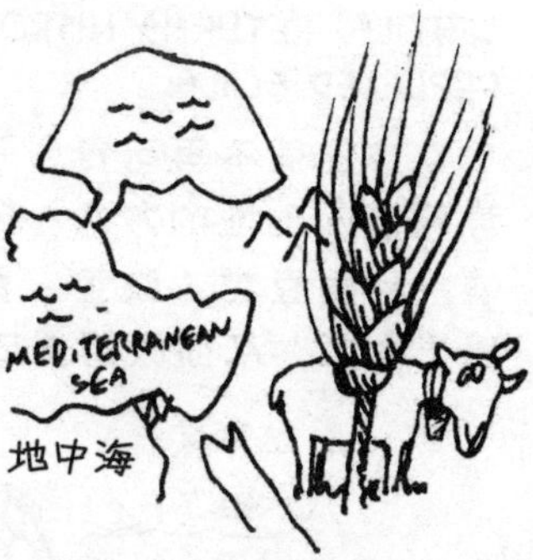

SOMEWHAT LATER, THE CHINESE MASTERED MILLET, PIGS, POULTRY, AND LATER STILL, RICE.

随后，中国人掌握了驯化小米、猪和鸡的技术，再后来，又能种植大米了。

MEXICO AND ITS NEIGHBORS, GROWING ***CORN*** (MAIZE), WERE HANDICAPPED BY THE PLANT'S GENETIC PECULIARITIES FROM BREEDING HIGH-YIELD VARIETIES UNTIL 2000 B.C. MEANWHILE, THEY BRED PEPPERS, SQUASH, TOMATOES, AND CHOCOLATE, WHILE THE PERUVIANS PLANTED POTATOES AND LOADED LLAMAS...

墨西哥及其邻国种植**玉米**。由于玉米的遗传特征的阻碍，直到公元前2000年，他们才培育出高产品种。同时，他们还种植辣椒、南瓜、西红柿和可可。而秘鲁人则种植马铃薯和饲养美洲驼。

WHILE AFRICANS PLANTED YAMS AND SORGHUM AND RAISED CATTLE.

非洲人种植山药和高粱，并且也养殖牛群。

BY ***2000*** B.C., VIRTUALLY EVERYTHING WE NOW FARM HAD BEEN DOMESTICATED. ONLY ABOUT ***50 ANIMAL SPECIES*** HAVE EVER BEEN TAMED (THE OTHERS CAN'T STAND IT), AND EVEN TODAY THE BULK OF THE WORLD RELIES ON A VERY LIMITED NUMBER OF STAPLE VEGETABLE FOODS.

到公元前2000年，我们的祖先几乎种植了我们现在种植的所有作物。但大概只有50**种动物**被驯养（其他的动物不能够忍受驯养的过程），甚至在今天，世界上很大一部分人只依赖有限的几种作物作为主食。

WHEREVER PEOPLE FARMED, THEY ENCOUNTERED SOME UNFORESEEN PROBLEMS.

只要种地，人们就会面临一些不可预知的问题。

THE MOST BASIC PROBLEM IS THAT FARMING IS ***EXTRACTIVE:*** NUTRIENTS COME OUT OF THE SOIL BUT DON'T CYCLE BACK, AS IN WILD ECOSYSTEMS.

最基本的问题就是：农业是**提取性质的**——正如在野生生态系统中一样，营养从土壤中被提取，但不能重新回到土壤中。

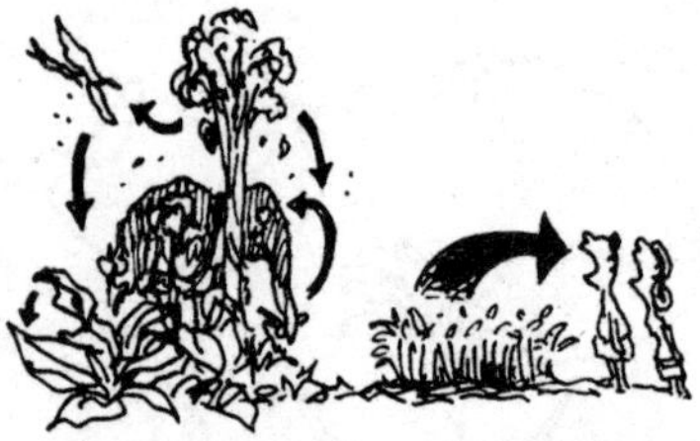

THROUGHOUT HISTORY, FARMERS HAVE DEALT WITH SOIL DEPLETION. THE ORIGINAL WAY WAS SIMPLY TO MOVE TO GREENER PASTURES.

纵观历史，古代农民也应对过土壤贫瘠的问题。原始的方法是：干脆迁移到更绿的草原去！

WHEN THAT BECAME IMPRACTICAL, THEY FOUND WAYS TO PUT NUTRIENTS INTO THE LAND: SPREADING MANURE RETURNS PHOSPHORUS, AND PLANTING LEGUMES (PEAS, BEANS, LENTILS) RETURNS NITROGEN TO DEPLETED SOILS.

当迁徙不再可行，于是，他们找到了将营养植入土地的方法：给土壤施粪肥会产生磷，种植豆类（豌豆、蚕豆、兵豆）会使土壤产生氮，从而改善贫瘠的土壤。

ANOTHER TACTIC IS ***INTER-PLANTING:*** GROWING DIFFERENT SPECIES OR VARIETIES IN THE SAME PLOT. IN CENTRAL AMERICA, THE COMBINATION WAS SQUASH, CORN, AND BEANS.

另一个策略就是**混合栽种**：即在同一块土地上种植不同的植物，或同一植物的不同品种。在中美，往往将是南瓜、玉米和大豆间种。

WHATEVER THE TECHNIQUE, SUCCESSFUL FARMING DEPENDS IN THE FIRST PLACE ON GOOD SOIL MANAGEMENT.

不管是什么技术，成功的农业首先依赖于良好的土壤管理。

ANOTHER AGRICULTURAL RISK IS EROSION, OR LOSS OF TOPSOIL.

农业的另一个风险就是侵蚀，或者说是表土层的流失。

IN WILD ECOSYSTEMS, MANY PLANTS, AND ESPECIALLY TREES, COMBINE TO HOLD WATER, PRODUCE AND ENRICH SOIL, AND ANCHOR IT WITH THEIR ROOTS.

在野生生态系统中，许多植物，尤其是树木交织在一起，从而保存水分，产生土壤，给土壤提供养料，并且用根将土壤固定。

CLEARED LAND DRIES UP AND IS EASILY BLOWN AWAY BY WIND OR WASHED OFF BY FLOODS.

空旷的土地干涸，土很容易被风或洪水冲走。

THIS IS WHY CAREFUL HILLSIDE FARMERS BUILD ***TERRACED FIELDS*** WATERED BY CONTROLLABLE DRAINAGE SYSTEMS.

这就是山区细心的农民为何开辟**梯田**的原因。他们通过可控排水系统来给土壤浇水。

DESPITE EVERYONE'S BEST EFFORTS, FARMING COMMUNITIES CAN STILL DEGRADE THEIR ENVIRONMENTS IN THE LONG RUN. WE OFFER THREE EXAMPLES, TWO ANCIENT AND ONE MODERN.

尽管每个人都付出了努力，但从长远来看，自耕农还是会破坏环境的。我们现在举三个例子，两个古代的和一个现代的。

SALTY SOIL STUNTS PLANTS, REDUCES YIELDS, AND EVENTUALLY MAKES A FIELD INFERTILE.

含盐土壤会阻碍植物的生长，降低植物产量，并最终导致耕地贫瘠。

UNLESS FIELDS ARE WELL FLUSHED FROM TIME TO TIME, SALT BUILDS UP IN THE SOIL.

如果土壤不能经常被冲刷，盐就会不断地在土壤中富集。

AT FIRST, SUMERIAN FARMERS SIMPLY ABANDONED SALTY PLOTS AND PLOWED ELSEWHERE...

起初，苏美尔农民干脆离开盐碱土地，转而去其他地方耕作……

BUT AS POPULATION SWELLED, IRRIGATION WORKS COVERED THE LAND. DIVERTING WATER TO FLUSH OUT SALTY FIELDS WOULD HAVE DISPLACED TOO MANY PEOPLE.

但是随着人口膨胀，灌溉工程覆盖了全部土地，因灌溉而造成的盐碱化土地导致大量人口流离失所。

AFTER MORE THAN ***2000 SUCCESSFUL YEARS*** OF FERTILITY, SUMER'S SOIL BECAME ALMOST USELESS. CROP YIELDS PLUNGED, AND BY 1700 B.C., POPULATION HAD FALLEN BY A FACTOR OF 10, TO AROUND 150,000.

2000**多年**丰饶的光辉岁月后，苏美尔的土地几乎没有任何用处了。粮食产量猛跌，到公元前1700年，人口数量已下降为原来的十分之一，只剩下大约15万。

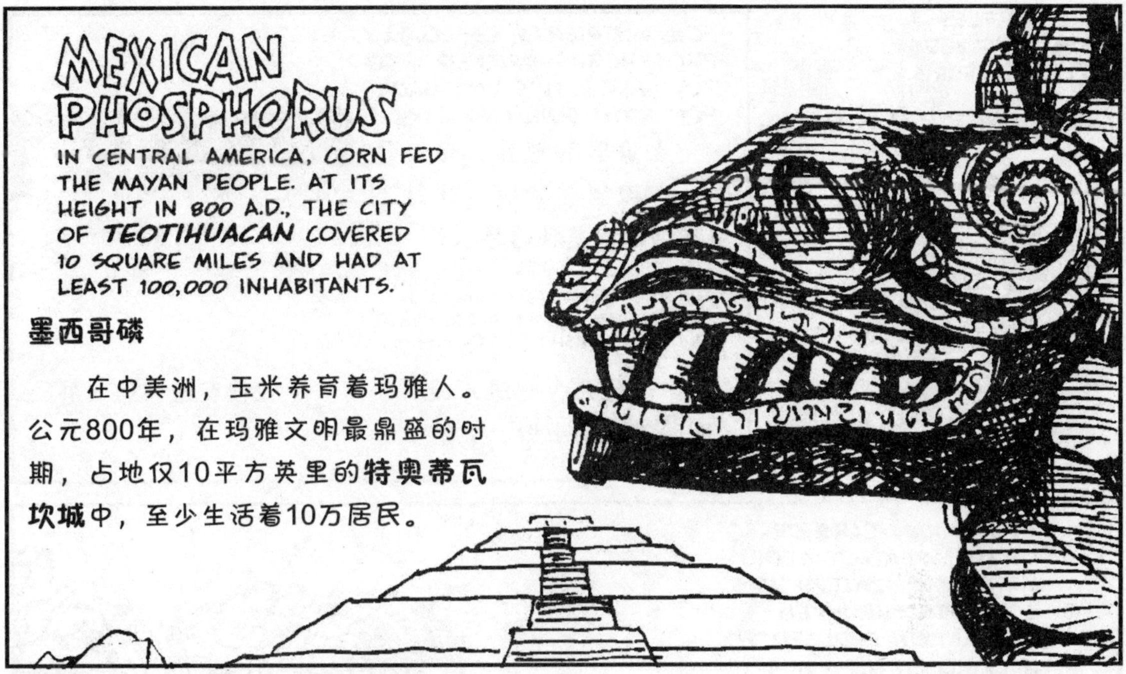
MEXICAN PHOSPHORUS
IN CENTRAL AMERICA, CORN FED THE MAYAN PEOPLE. AT ITS HEIGHT IN 800 A.D., THE CITY OF *TEOTIHUACAN* COVERED 10 SQUARE MILES AND HAD AT LEAST 100,000 INHABITANTS.
墨西哥磷
在中美洲，玉米养育着玛雅人。公元800年，在玛雅文明最鼎盛的时期，占地仅10平方英里的**特奥蒂瓦坎城**中，至少生活着10万居民。

BUT THE SEEDS OF ITS RUIN WERE ALREADY SOWN. THE DEFORESTATION OF MEXICO WAS COMPLETED AROUND THE YEAR 250.
但消亡的种子已被悄然播种。大约公元250年，墨西哥的森林彻底被破坏。
玉米长得真高啊！
THE CORN IS AS *HIGH*... AS... AS... A...

DESPITE ALL THE MAYAS' EFFORTS AT WATER CONTROL, TERRACING, AND RAISED FIELDS (FOR BETTER DRAINAGE), EROSION WENT STEADILY ON.
尽管玛雅人在控制水源、修梯田、垫高土壤（有利于排水）等方面付出了很多努力，但是土壤还是不断地被侵蚀。
就像一只小鸡的喉咙？
AS A CHICKEN'S LARYNX?
PUK
咕咕

CORE SAMPLES FROM LAKE BEDS TELL THE STORY:

取自湖床的矿样告诉我们：

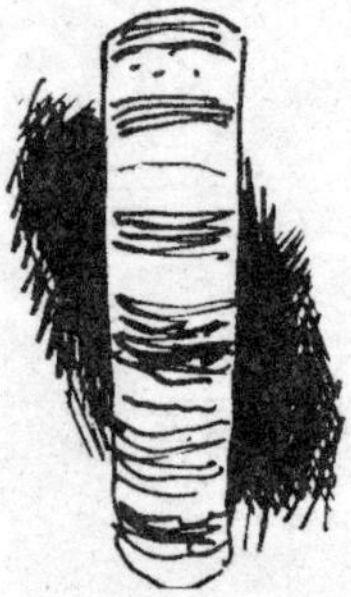

SOIL NUTRIENTS, ESPECIALLY PHOSPHORUS, DRAINED INTO THE LAKES. THIS WAS BAD FOR BOTH SOIL AND LAKE.

土壤里的营养，尤其是磷，都流到了湖里。这对土壤和湖都不是件好事。

(AND THE MAYAN PEOPLE HAD NO DOMESTICATED ANIMALS WHOSE MANURE COULD REPLENISH THE POOR SOIL.)

（另外，玛雅人没有家畜，也就没有肥料可以补充给贫瘠的土壤。）

WITHIN 100 YEARS OR SO OF ITS GREATEST GLORY, MAYAN CIVILIZATION WAS GONE, ITS SHRUNKEN POPULATION REDUCED TO HACKING LITTLE FARM PLOTS OUT OF THE JUNGLE.

在到达辉煌顶峰之后的不到100年，玛雅文明便一去不复返。稀落的人们沦落到在丛林中开辟小片荒地来耕作的地步。

SILT & DE-NILE

SINCE ANCIENT TIMES, THE NILE RIVER HAS OVERFLOWED ITS BANKS IN EGYPT EVERY YEAR.

尼罗河的淤泥

自古以来，尼罗河在埃及每年都会泛滥。

IN THIS WAY, A FRESH LOAD OF FERTILE SILT IS REPEATEDLY DROPPED ON EGYPTIAN FARMLANDS.

这么一来，肥沃的淤泥不断地停留在埃及的农田上。

THE SILT, IN TURN, COMES FROM THE HIGHLANDS OF UGANDA AND ETHIOPIA, THE SOURCES OF THE NILE.

那些淤泥来源于乌干达和埃塞俄比亚的高地，而那里正是尼罗河的发源地。

UNFORTUNATELY, THIS SYSTEM WAS INCONVENIENTLY UNPREDICTABLE. SOMETIMES THE FLOODS WERE TOO LITTLE—OR TOO MUCH.

不幸的是，这个系统很不方便，且不可预测。洪水时而太少，时而太多。

IN THE 1960S, THE EGYPTIAN GOVERNMENT ***DAMMED*** THE NILE AT ***ASWAN*** TO CONTROL FLOODS, GENERATE ELECTRICITY, AND INCREASE THE ACREAGE UNDER IRRIGATION.

在20世纪60年代，埃及政府修建了**阿斯旺大坝**来控制洪水和发电，增加可灌溉土地面积。

NOW, INSTEAD OF FERTILIZING THE NILE'S BANKS, THE SILT SETTLES OUT IN THE LAKE BEHIND THE DAM… DOWNSTREAM FARMERS NEED CHEMICAL FERTILIZERS… NILE FISH STARVE… SNAILS SPAWN, SPREADING THE AWFUL DISEASE CALLED SCHISTOSOMIASIS…

现在，那些淤泥沉淀在水坝下面的湖中，而不是给尼罗河河岸的土地提供肥料。下游的农民需要化学肥料；尼罗河的鱼群被饿死；蜗牛大量产卵，传播一种叫作血吸虫病的疾病，太可怕了……

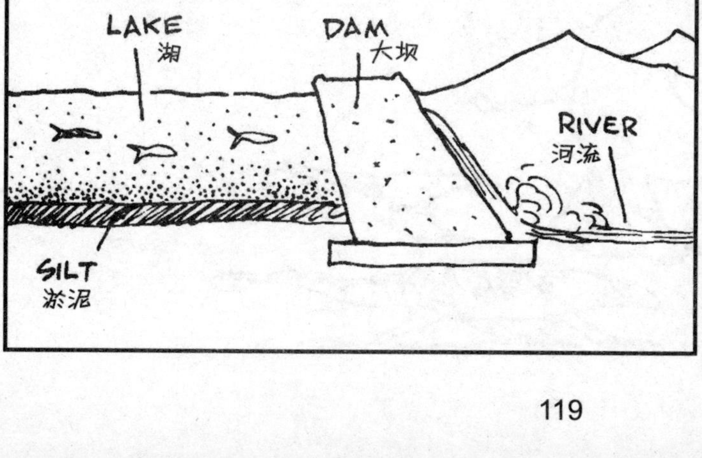

AND THE DAM'S ***HYDRO-ELECTRIC POWER*** IS MAINLY USED TO RUN THE FERTILIZER FACTORIES!

大坝发的**水电**主要用来为肥料厂提供动力。

DESPITE ALL ITS PROBLEMS, AGRICULTURE HAS SPREAD WORLDWIDE, SO THAT TODAY ***99.9%*** OF THE WORLD'S PEOPLE DEPEND ON FARMING FOR FOOD. BY DIVERTING ENORMOUS AMOUNTS OF BIOLOGICAL ENERGY TO ITS OWN BENEFIT, THE HUMAN POPULATION HAS SWELLED FROM A FEW MILLION IN *10,000* B.C. TO AT LEAST ***5 BILLION*** TODAY, AND COUNTING.

尽管存在很多问题，但农业还是在全球传播开来，以至于当今世界99.9%的人口要依靠农业生产的食物。通过将大量的生物能转化为自身所需，人类的数量已经从公元前1万年的几百万膨胀到今天的至少50亿*。

* 到2011年全球人口已达70亿。

THE HAPLESS ORGANISMS ON THE LOSING END OF THIS PROPOSITION MIGHT WELL WONDER: DON'T HUMAN POPULATION HAVE LIMITING FACTORS TOO??

在整个过程中，那些与人类竞争失败的不幸生物可能会想：难道人类就没有控制其数量的限制因子？

CHAPTER 8

WHAT LIMITING FACTORS?

第八章

有什么限制因子啊？

HOW DO POPULATIONS GROW?
WHY ARE THERE SO MANY PEOPLE?
AND WHY AREN'T THERE EVEN MORE?

人口是怎样增长的？
为什么现在有这么多人？
又为何不是更多呢？

BEFORE EXPLORING THE EXPANSION OF THE HUMAN SPECIES, CONSIDER THIS PARADOX: AFTER THE INVENTION OF AGRICULTURE, MOST PEOPLE WERE ***WORSE OFF*** THAN THEIR ANCESTORS. FROM SKELETONS WE KNOW THAT WHEN AGRICULTURE ARRIVED, AVERAGE HEIGHT DECLINED ABOUT 4 INCHES, FROM 5'9" TO 5'5" FOR MEN AND FROM 5'5" TO 5'1" FOR WOMEN. COMPARED TO HUNTER-GATHERERS, FARMERS WORK HARDER, EAT A LESS NUTRITIOUS DIET, AND GET SICK MORE OFTEN.

在探索人口膨胀之前，首先让我们来考虑一下这个相互矛盾的现象：人类在发明了农业之后，大多数人的生活反而不如他们的祖先丰实。从对骨骼的研究来看，我们发现，随着农业的到来，人类的平均身高下降了4英寸*，男性的平均身高从5.9英尺下降到5.5英尺，女性的平均身高从5.5英尺下降到5.1英尺。与狩猎者和采集者相比，农业生产者的劳动更加辛苦，食谱不如以前有营养，也更容易生病。

*1英寸=0.0254米，下同。

A SYSTEM, AGRICULTURE, THAT MAKES PEOPLE SMALL AND SICKLY, HAS SUPPLANTED A SYSTEM, HUNTING AND GATHERING, THAT MAKES PEOPLE HEALTHY AND ROBUST. HOW CAN THIS BE?

农业系统使得人类变得矮小和体弱多病，狩猎—采集系统使人类更加健康和强壮。可最终农业系统却取代了狩猎—采集系统，这是怎么回事呢？

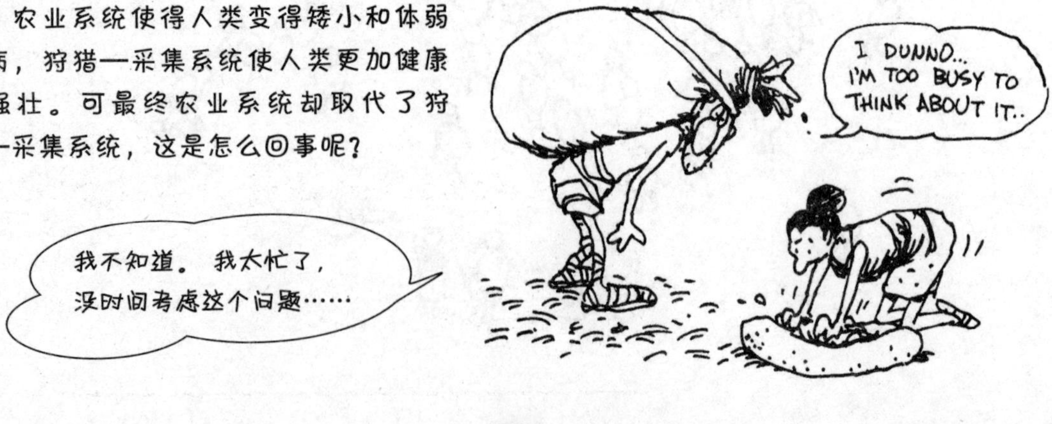

THE ANSWER LIES IN AGRICULTURE'S ***SYSTEM DYNAMICS:*** ONCE IT BEGINS, FARMING TENDS TO EXPAND!

答案要从农业的**系统动力学**里去找。农业一旦开始，定会扩大。

IN THE FIRST PLACE, AGRICULTURE GENERALLY PRODUCES A ***SURPLUS:*** AT HARVEST TIME, THERE ARE HEAPS OF GRAIN OR GOURDS OR POTATOES—TOO MUCH TO EAT ON THE SPOT, SO IT'S PUT INTO STORAGE FOR THE WINTER. MOST OF THE COMMUNITY'S FOOD IS IN ONE PLACE AT ONE TIME.

首先，农业生产的东西通常都会有**剩余**的：在收获季节，有成堆的谷物和满地的瓜类或土豆——实在太多，一时是吃不完的，所以得储存起来冬天吃。一个群体的大部分食物每次只储存在一个地方。

NOW THE QUESTION ARISES, WHO OWNS THAT SURPLUS?

现在问题出现了，谁是那些剩余粮食的主人呢？

UNLIKE HUNTING AND GATHERING, AGRICULTURE ENABLES THE STRONGEST PEOPLE IN THE COMMUNITY TO LAY HANDS ON EVERYONE'S ENERGY RESOURCES AND CONTROL THEM.

与狩猎和采集不同的是，农业使得群体中最强壮的人向每个人索要能源，并掌握控制权。

IN OTHER WORDS, WITH AGRICULTURE COMES A ***LANDLORD CLASS:*** KINGS, PRIESTS, AND NOBLES, WHO TAKE A CUT OF THE HARVEST FOR THEMSELVES, AND NOT A SMALL ONE EITHER—50% IS TYPICAL!

换句话说，农业使**地主阶级**产生了，即国王、僧侣、贵族等，他们开始搜刮贫民的丰收成果，而且一般都要征收50%，这可不是个小数目！

THAT CONCENTRATED SURPLUS IS THE SOURCE OF ***SOCIAL ORGANIZATION.*** THE LANDLORDS USE THEIR WEALTH TO HIRE ARMIES, BUILD ROADS, AND HIRE THINKERS TO DREAM UP IDEAS TO KEEP THE PEASANTS IN LINE.

剩余产品的集中是产生**社会组织**的根源。地主利用他们的财富雇佣军队，修建道路，还雇佣思想家献计献策，以便控制农民。

THE WHOLE POINT OF THIS EXERCISE CALLED CIVILIZATION IS TO KEEP THE SURPLUSES OF ENERGY—STARTING WITH FOOD ENERGY—FLOWING INTO THE HANDS OF THE STRONG. IN OTHER WORDS, IT WAS ORGANIZED FROM THE OUTSET TO BE ***SELF-PERPETUATING.***

所谓文明，其全部含义就是将剩余的能量——从食物能量开始，掌握在强者手中。换句话说，从一开始其组织模式就是为了达到**自身永存**的目的。

THEN, TOO, WE'VE SEEN HOW THE LOGIC OF FARMING COMPELS THE POOR TO HAVE MORE CHILDREN RATHER THAN FEWER: MORE HANDS MEANS MORE WORK CAN BE DONE.

同样，我们已经明白为什么农业生产的结果是促使穷人生育更多的孩子，因为人手越多，壮劳力越多。

AND IF YOU SHOULD WANT TO STOP WORKING AND GO BACK TO HUNTING, FORGET IT! HUNTING WAS STRICTLY REGULATED, EVEN FORBIDDEN TO COMMONERS. THIS IS A SPORT FOR THE NOBLES!

如果你不想干活想回到狩猎时代，休想！狩猎是被严格控制的，普通人甚至被排斥在外，狩猎可是一项贵族运动！

THE CONSEQUENCE IS ***RAPID POPULATION GROWTH.*** THE NEW GENERATIONS CLAIM MORE LAND FOR FARMING... TRADITIONAL PEOPLES, WITH THEIR POPULATION CONTROLS, ARE OUTNUMBERED AND OVERPOWERED... NO WONDER AGRICULTURE TOOK OVER THE WORLD!

后果便是**人口迅猛增长**。新一代人需要更多的土地来耕作……而持有传统观念控制人口增长的人，也就随之淡出历史舞台，被排挤走了。难怪农业统治了整个世界。

IN GENERAL, ANY POPULATION OF ORGANISMS, IF UNCHECKED, TENDS TO GROW ***EXPONENTIALLY.*** WHAT DOES THAT MEAN?

总的来说，任何生物体的数量如果不加以控制，一般都会呈**指数**增长。这是什么意思呢?

IT MEANS THAT THE POPULATION'S ***RATE OF INCREASE*** IS ***CONSTANT.*** IN SYMBOLIC TERMS, WE CALL P_0 THE POPULATION TODAY AND r THE RATE OF INCREASE. THEN A YEAR FROM NOW THE POPULATION WILL BE $P_0 + rP_0$. A YEAR LATER, WITH r REMAINING THE SAME, THE POPULATION WILL BE $(P_0 + rP_0) + r(P_0 + rP_0)$. AFTER A TINY BIT OF ALGEBRA, THIS BECOMES $P_0(1+r)^2$, AND WE CAN GO ON TO WRITE:

意思是**人口增长率是恒定**的。可以用符号来表示，我们用P_0表示当今人口数，r表示增长率。从现在算一年后，人口将会是P_0+rP_0。再过一年后，r不变，人口是（P_0+rP_0）+r（P_0+rP_0）。掌握一点代数知识，便可以写成P_0（$1+r$）^{2}。我们可以继续计算：

POPULATION TODAY........ P_0

" AFTER **1** YEAR..... $P_0(1+r)$

" " **2** YEARS.... $P_0(1+r)^2$

" " **3** YEARS.... $P_0(1+r)^3$

⋮

" " **n** " .... $P_0(1+r)^n$

当今人口……P_0

一年后……P_0（$1+r$）

两年后……P_0（$1+r$）^{2}

三年后……P_0（$1+r$）^{3}

⋮

N年后……P_0（$1+r$）^{n}

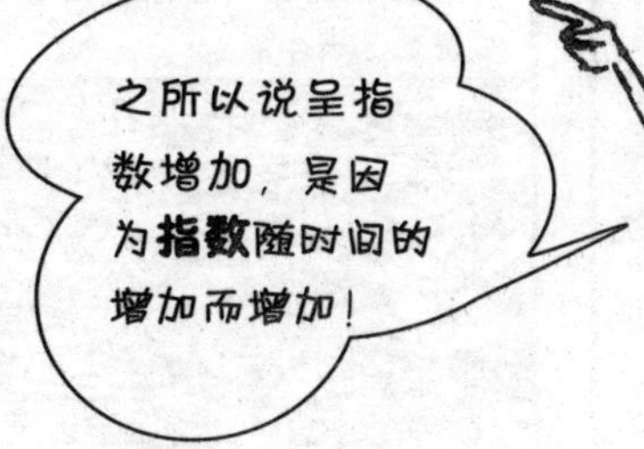

DRAW A GRAPH OF THIS ***EXPONENTIAL FUNCTION***, AND YOU SEE THAT ITS LEFT END LOOKS QUITE DIFFERENT FROM ITS RIGHT END. AT FIRST IT GROWS ONLY SLOWLY, ALMOST IMPERCEPTIBLY, BUT AT SOME POINT IT TAKES OFF. (HERE ARE TWO EXAMPLES, ONE WITH LOW r AND ONE WITH HIGH r.)

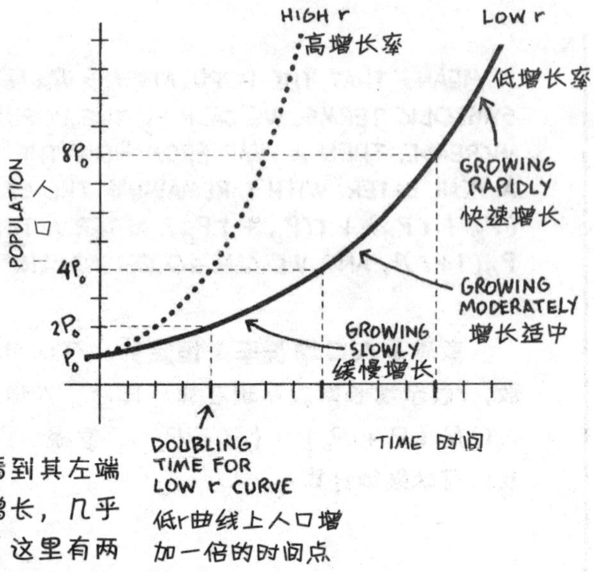

这种**指数**可以用曲线图表示，你可以看到其左端与右端看起来大不相同。起初它很缓慢地增长，几乎觉察不到，但从某一个点开始快速增长。（这里有两个例子，一个是低增长率，一个是高增长率。）

IF YOU AREN'T COMFORTABLE WITH THE ALGEBRA, ALL YOU NEED TO REMEMBER IS THIS: EXPONENTIAL GROWTH HAS A FIXED

DOUBLING TIME.

THERE IS A CERTAIN NUMBER OF DAYS, OR MONTHS, OR YEARS (IT DEPENDS ON THE SPECIES) IN WHICH THE POPULATION DOUBLES.

如果你对代数不怎么感兴趣，你只需要记住：呈指数增长总有一个固定的**倍增时间**。这个点可能是数天之后、数月之后或者数年之后（这取决于物种种类），在此时段人口会增长一倍。

LIKE THE WATER LILIES, NO POPULATION CAN CONTINUE EXPONENTIAL GROWTH FOREVER: RESOURCES ARE FINITE, AND BIOGEOCHEMICAL CYCLES ARE TOO SLOW TO SUPPLY FRESH RESOURCES TO A RAPIDLY GROWING POPULATION.

看了睡莲的例子后，你就知道了人口也不能永远持续呈指数增长，因为资源是有限的，生物地球化学循环的速度太慢，无法为迅速增长的人口提供新鲜资源。

WHAT HAPPENS WHEN A GROWING POPULATION HITS THE CARRYING CAPACITY OF ITS ENVIRONMENT?

当不断增长的人口达到环境的承载力时，将会发生什么？

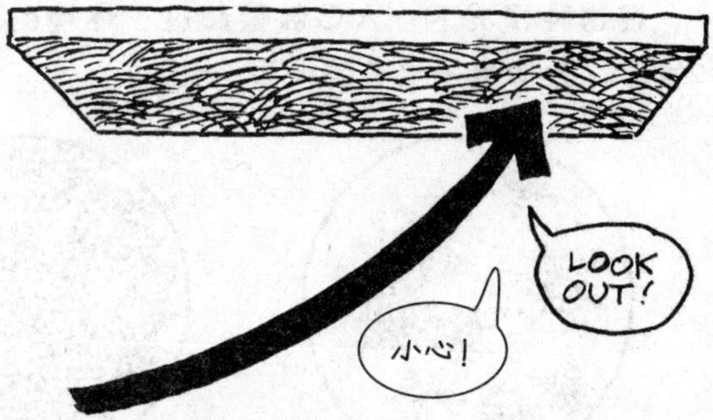

WITH ANY LUCK, IT WILL SLOW AS IT APPROACHES THE CARRYING CAPACITY AND LEVEL OFF, REACHING A HAPPY ACCOMMODATION WITH THE ENVIRONMENT. THIS GIVES AN **S**-CURVE.

如果幸运的话，人口在接近环境的承载能力时，增长速度会放缓，并趋于稳定，与环境形成和谐关系。这就形成一个**S**曲线。

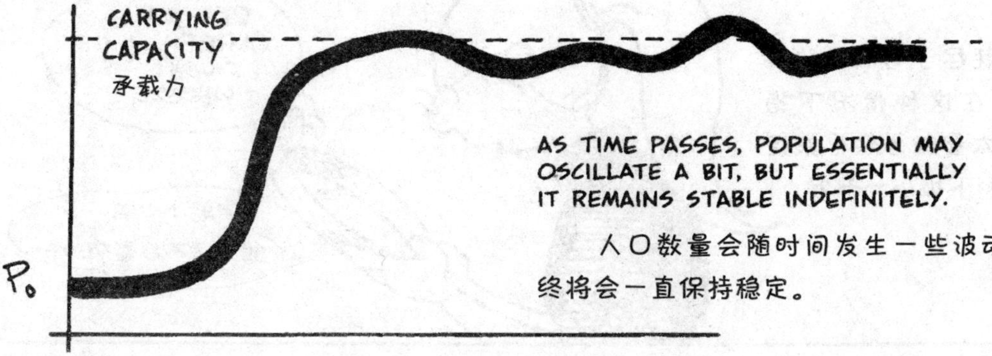

AS TIME PASSES, POPULATION MAY OSCILLATE A BIT, BUT ESSENTIALLY IT REMAINS STABLE INDEFINITELY.

人口数量会随时间发生一些波动，但最终将会一直保持稳定。

A LESS ROSY SCENARIO IS DESCRIBED BY THE

J-CURVE,

WHICH LOOKS LIKE THIS: A PERIOD OF RAPID POPULATION INCREASE FOLLOWED BY COLLAPSE.

另一个不这么乐观的情况可以用J曲线来表现，情况大致是这样的：人口快速增长一个阶段后，便立即暴跌。

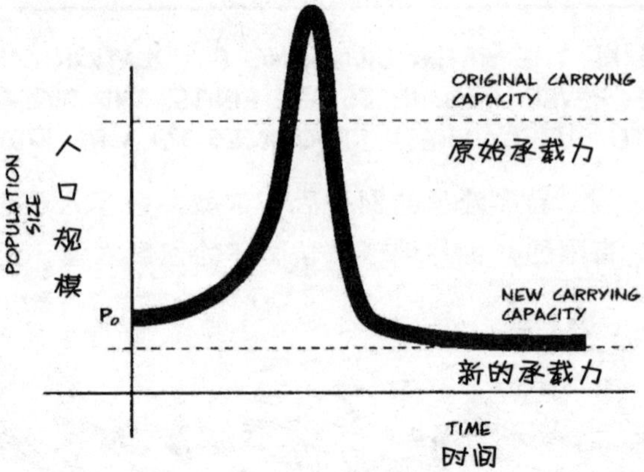

IN THIS CASE, A POPULATION OVERSHOOTS THE ENVIRONMENT'S CARRYING CAPACITY, AND THE SPECIES EATS ITSELF OUT OF HOUSE AND HOME, LIKE A BACTERIAL CULTURE ON A PETRIE DISH.

在这种情况下，人口数量超过了环境的承载力，人类便会将自己挤出住所和家园，就像盘子里养的细菌一样。

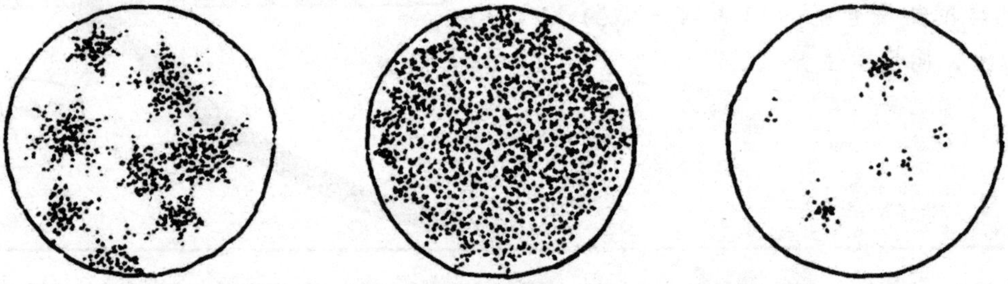

AFTER DEPLETING THE AVAILABLE RESOURCES (SUGAR IN THIS CASE), THE BACTERIA DIE OFF RAPIDLY, LEAVING ONLY A PATHETIC REMNANT OF THE ORIGINAL POPULATION.

在耗尽了可利用的资源后（在这种情况下指糖），大量细菌会很快死亡，只剩下极小一部分。

WHEN IT COMES TO THE STUDY OF HUMAN POPULATION, THE FIRST NAME HAS TO BE THOMAS

MALTHUS

(1766-1834), FAMOUS PESSIMIST.

讲到对人类人口的研究，首先必须提到**托马斯·马尔萨斯**（1766—1834），他是著名的悲观主义者。

MALTHUS IS BEST KNOWN FOR SAYING THAT FOOD PRODUCTION CAN RISE ONLY ***LINEARLY***—I.E., WHEN YOU GRAPH FOOD AGAINST TIME, YOU GET A STRAIGHT LINE. (THE REASON IT GROWS AT ALL, HE THOUGHT, WAS THE CULTIVATION OF NEW FARMLAND.)

马尔萨斯的著名论点是：食物生产只能以**线性**速度增长——如果将食物生产和时间的关系用图来表示，得到的一定是一条直线。（他认为食物生产增加的原因是开垦了新的耕地）。

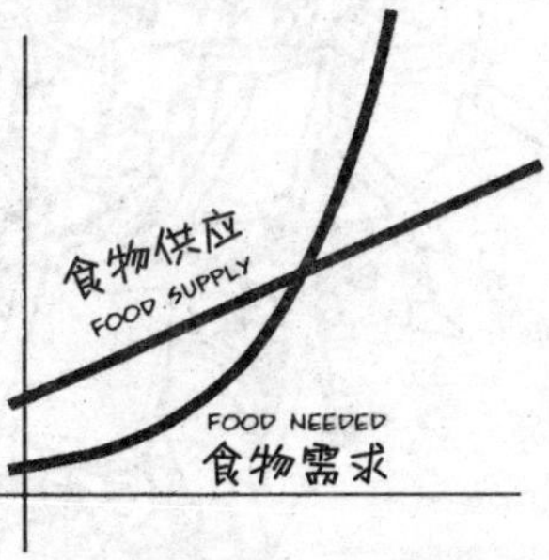

INEVITABLY, HE SAID, EXPONENTIALLY GROWING POPULATION WILL OUTSTRIP THE AVAILABLE FOOD AT SOME POINT...

他指出，呈指数增长的人口数有朝一日将不可避免地超过食物供应量。

BUT MALTHUS WAS NOT COMPLETELY GLOOMY! HE OPTIMISTICALLY MAINTAINED THAT HUMAN NUMBERS WERE HELD IN BALANCE WITH THE FOOD SUPPLY BY THREE "PREVENTATIVE CHECKS."

但是马尔萨斯并不完全悲观！他认为，有三种原因可以使人口和食物供给保持平衡，那就是

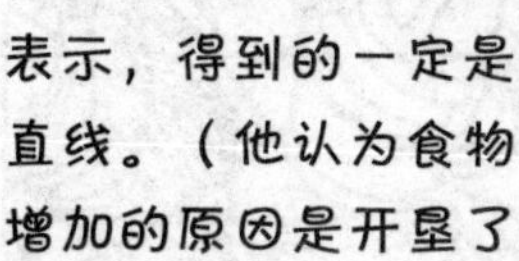

PESTILENCE, FAMINE, AND WAR.

瘟疫、饥荒、战争。

IT TURNS OUT THAT MALTHUS WAS WRONG ABOUT THE FOOD SUPPLY. FOR REASONS WE'LL DISCUSS IN A LATER CHAPTER, FARMERS HAVE INCREASED THEIR TOTAL OUTPUT AT A MUCH FASTER RATE THAN A STRAIGHT LINE.

HE WAS RIGHT, HOWEVER, TO IDENTIFY PESTILENCE, FAMINE, AND WAR AS FACTORS THAT LIMIT HUMAN POPULATIONS... SO LET'S LOOK AT EACH IN TURN.

事实证明，马尔萨斯有关食物供给的观点是错误的。至于原因我们将会在下一章中进行讨论，农业生产者以较快的速度而不是以直线增长的速度增加总产量。

然而，他认为瘟疫、饥荒和战争是限制人类人口数量增长因素的观点却是正确的。那么，我们就分别来研究一下吧！

WAR

HAS BEEN WITH US SINCE THE DAWN OF HISTORY. IT REDUCES POPULATION IN AN OBVIOUS AND DIRECT WAY: BY KILLING PEOPLE.

纵观历史，自从文明的曙光照耀大地，**战争**就一直伴随着我们。它以一种明显而直接的方式，即屠杀，导致人口数量减少。

BUT GROUPS WAGE WARS TO WREST RESOURCES FROM EACH OTHER. THE WINNING SIDE, DESPITE ITS BATTLE LOSSES, MAY GAIN ENOUGH TO THRIVE AFTERWARD, INCREASING POPULATION.

人类群体之间发动战争是为了相互争夺资源。尽管获胜的一方在战争中有损失，但也可能会为日后的繁荣获取足够的资源，这样人口也会重新增加。

BESIDES, BIRTHRATES USUALLY GO UP AFTER WARS, WHEN SOLDIERS COME HOME TO THEIR WIVES, AND POPULATIONS REBOUND RAPIDLY.

此外，当战后士兵回家与妻子团聚后，人口出生率通常会增加，使得人口快速反弹。

WAR DOES MOST OF ITS DAMAGE INDIRECTLY, BY SPREADING DISEASE AND FAMINE.

战争所带来的破坏大部分是间接的，如疾病传播和饥荒便是间接的破坏。

AN ARMY MARCHES ON ITS STOMACH, AND IN THE DAYS BEFORE CANNED AND FREEZE-DRIED RATIONS, THIS MEANT STRIPPING THE COUNTRYSIDE.

行军专靠肚皮饱。但在罐头食品和冷冻食品没有被发明之前，且无法被定额发放的日子里，这意味着军队会将沿途的村庄洗劫一空。

ABSENT AND DEAD SOLDIERS AREN'T AVAILABLE FOR FIELD WORK, EITHER.

逃兵和死去的士兵也不能干农活。

FOOD SHORTAGES LEAVE PEOPLE SUSCEPTIBLE TO DISEASE, AND ARMIES OFTEN BRING HOME FOREIGN INFECTIONS.

食物短缺使得人们易于得病，而且军队经常会将国外的传染病带回国内。

OF COURSE, ONE MAY PERFECTLY WELL STARVE WITHOUT A WAR.

FAMINE

HAS BEEN A PART OF AGRICULTURE FROM EARLY TIMES.

当然，在没有战争时人们也有挨饿的时候。**饥荒**从很早的时候就成了农业社会中的一部分。

EVEN IN PEACETIME, FARMERS HAVE BEEN BADLY NOURISHED COMPARED TO THEIR HUNTER-GATHERER COUSINS. WHEN THE HARVEST IS BAD, THE EFFECT IS MALNUTRITION OR STARVATION.

甚至在和平年代，农民的营养水平也不如狩猎者和采集者。收成不好时，后果就是营养不良或饥荒。

(BY CONTRAST, THE HUNTER-GATHERER, WITH A WIDER VARIETY OF FOOD SOURCES, IS BETTER ABLE TO SURVIVE THE LOSS OF ONE OR MORE OF THEM.)

（相反，狩猎者和采集者的食物来源众多，失去了一种食物，还可以吃其他的，因此更有机会存活下来。）

BEFORE THE ADVENT OF HIGH-SPEED TRANSPORT AND INTERNATIONAL RELIEF AGENCIES, EVERY LOCAL CROP FAILURE MEANT FAMINE. CHINESE OFFICIALS RECORDED 1,828 FAMINES BETWEEN THE YEARS 108 B.C. AND 1910. A BRITISH LIST RECOUNTS 11 FAMINES IN THE 13TH CENTURY ALONE.

在高速交通和国际救援时代到来之前，每当一个地方作物歉收时，都会发生饥荒。据记载，中国在公元前108年到1910年间发生了1828次饥荒。据统计，英国单在13世纪就发生了11次饥荒。

FAMINES TEND TO BE LOCAL: A POOR HARVEST IN ONE REGION HAS LITTLE EFFECT ON FOOD SUPPLIES ELSEWHERE. THEREFORE, THE EFFECT OF FAMINE ON GLOBAL POPULATION IS MINOR, DESPITE THE DEVASTATION IN THE IMMEDIATE NEIGHBORHOOD.

饥荒往往是地方性的，也就是说，一个地方收成不好很少会影响其他地方的食物供给。因此，虽然饥荒会对当地产生灾难性影响，但对全球人口的数量影响很小。

WITH PESTILENCE

IT'S DIFFERENT: DISEASES CAN SPREAD ACROSS CONTINENTS.

但是瘟疫的影响就大不一样了。瘟疫会跨洲越洋进行传播。

FOR EXAMPLE, IN THE 13TH CENTURY PLAGUE CAME TO EUROPE, CARRIED BY SAILORS ARRIVING FROM ASIA. BY THE END OF THE CENTURY, EUROPEAN POPULATION HAD DROPPED BY ABOUT HALF.

例如，13世纪的时候，航海家从亚洲归来，将鼠疫带到欧洲，到13世纪末，欧洲的人口减少了近一半。

TO UNDERSTAND EPIDEMIC DISEASES, YOU HAVE TO SEE THINGS FROM THE MICROBE'S POINT OF VIEW: THE BACTERIA, VIRUSES, AND AMOEBAS THAT CAUSE DISEASES ARE, LIKE OURSELVES, ORGANISMS LOOKING FOR A HABITAT.

为了更好地理解流行病，我们必须从微生物的角度来思考。引发疾病的细菌、病毒和变形虫像我们一样，也是要寻找到栖息地的生物体。

*病原：能引起疾病的微生物和寄生虫等病的统称。

TO SOME OF THESE PATHOGENS,* A WARM HUMAN BODY IS LIKE HEAVEN ON EARTH.

对于有些病原*来说，温暖的人体简直就是人间天堂。

BUT THESE LITTLE r-STRATEGISTS, BREEDING LIKE MAD, MAY ***KILL THEIR HOST***, AN UNFAVORABLE OUTCOME FOR THE GERMS, WHO ALSO SUFFER A DIE-OFF.

这些小东西利用自己的繁殖策略疯狂繁殖，可能会将其**寄主**害死。这对细菌来说也是不利的，因为失去了“家”，它们也会一同饿死。

IT'S MUCH BETTER FOR THE LITTLE INFECTIOUS AGENTS WHEN PLENTY OF HOSTS REMAIN ALIVE, AT LEAST IN SOME CONDITION.

如果大量的寄主活着，至少在一些情况下，这些小传染体会好一些。

THEY HAVE SEVERAL STRATEGIES TO ACCOMPLISH THAT END.

它们会采取几种策略以达到目的。

THEY MAY ***COEVOLVE*** WITH HUMANS. PEOPLE WHO SURVIVE THE DISEASE MAY BE GENETICALLY EQUIPPED WITH A MORE EFFECTIVE IMMUNE RESPONSE TO THAT PARTICULAR MICROBE...

它们可能会和人类**协同进化**。能够战胜疾病的人可能在遗传方面具有对这一特定微生物更有效的免疫力……

THE SURVIVING PATHOGEN ALSO MUTATES TO A LESS VIRULENT FORM.

幸存的病原体也会突变为一种毒性较弱的形式。

快走，女儿，要做一个善良、温柔的细菌！

嘿！嘿！

THE RESULT IS A MORE RESISTANT HUMAN POPULATION, INFECTED BY A LESS DEADLY FORM OF THE DISEASE. THE DISEASE IS NOW SAID TO BE ***ENDEMIC***.

结果产生了一群抵抗力更强的人类，他们只会染上一些毒性较弱的疾病。这种疾病叫作地方病。

THE SUCCESSFUL INFECTION IS ONE THAT ALLOWS ENOUGH HOSTS TO LIVE, WHILE AVOIDING DESTRUCTION BY THE HOST'S IMMUNE SYSTEM.

成功的疾病是指那些能够允许足够的寄主活下来，同时又不遭到寄主免疫系统破坏的疾病。

再见了，祖母！

THIS EXPLAINS WHY FOREIGN DISEASES CAN BE SO DEVASTATING TO ISOLATED POPULATIONS: IT TAKES TIME TO COEVOLVE. FOR EXAMPLE, WHEN EUROPEANS FIRST LANDED IN AMERICA, THEIR DISEASES KILLED FAR MORE NATIVE AMERICANS THAN EUROPEAN WEAPONS.

这就是为何外来的疾病会给不与外界接触的人群带来灾难性后果的原因。协同进化需要时间。例如，当欧洲人第一次登陆美洲时，他们带去的疾病比欧洲武器杀死的美洲当地人还要多。

EUROPE, MEANWHILE, HAD ALSO FALLEN VICTIM TO HORRIBLE PLAGUES. IN 430 B.C., 160 A.D., 540, THE 1300s, ETC. ETC. ETC. USUALLY, THESE WERE IMPORTED FROM ELSEWHERE BY SAILORS.

同时，欧洲也成了可怕的瘟疫的受害者。在公元前430年，公元160年、公元540年、公元1300年等，这些瘟疫通常都被航海者从其他地方带去。

EVENTUALLY, THE WORST PASSES, THE DISEASE BECOMES ENDEMIC, AND POPULATION AGAIN RISES STEADILY—UNTIL THE NEXT THING COMES AROUND.

最糟糕的时候终于过去了，疾病变成地方性的，人口数量再次开始稳步上升——直到下一次疾病的出现。

ANOTHER MICROBIAL STRATEGY IS TO SWITCH HOSTS. IF YOUR HUMAN DIES, GO ELSEWHERE!

细菌繁殖的另一种策略就是转变寄主。如果寄主死了，细菌就会转移到其他地方。

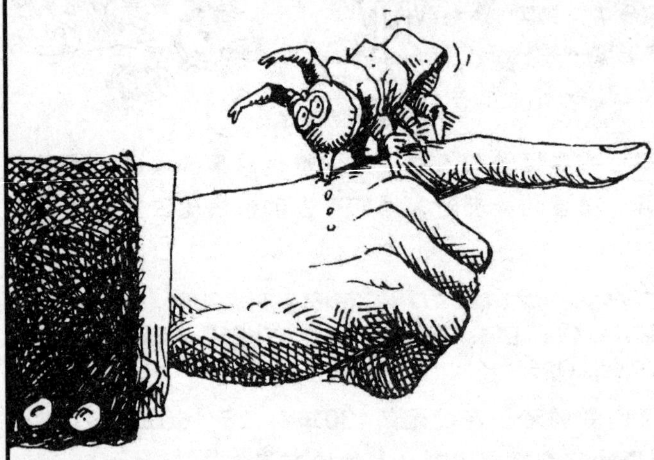

TYPHUS AND PLAGUE REALLY DO BEST ON FLEAS. THE FLEAS RIDE BLACK RATS INTO HUMAN HOUSEHOLDS, AND WHEN FLEA BITES HUMAN, THE INFECTION SPREADS.

斑疹伤寒和鼠疫最喜欢跳蚤。跳蚤随着黑家鼠来到人类的家里，一旦跳蚤咬人，传染病就开始传播。

INFLUENZA PREFERS PIGS AND DUCKS. CHINA, WHERE FARMERS RAISE PIGS AND DUCKS TOGETHER, IS THE BREEDING GROUND FOR NEW STRAINS OF FLU VIRUS, WHICH CROSSES THE PACIFIC OCEAN EVERY YEAR.

流感则更喜欢猪和鸭子。将猪和鸭子养在一起，会给流感病毒的新菌株提供温床，这些病毒甚至每年可以越洋传播。

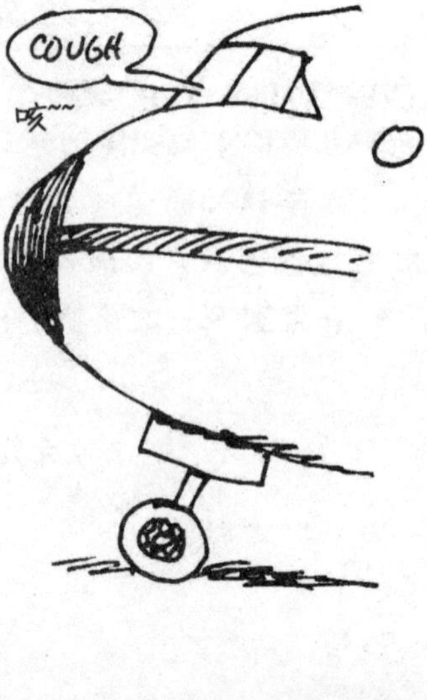

CONTACT BETWEEN CULTURES SPREADS DISEASE. THUS, EPIDEMICS HAVE OFTEN OCCURRED DURING PERIODS OF PEACE, WHEN DISTANT CIVILIZATIONS COULD SAFELY MAKE CONTACT BY SAIL OR CARAVAN. SIMULTANEOUS PLAGUES IN CHINA AND ROME IN THE 160S MAY HAVE RESULTED FROM TRADE BETWEEN THE TWO COUNTRIES.

文化交流也会传播疾病。因此，流行病经常发生在和平时期，此时远方文明之间可以安全地通过航行或商队相互接触。如在公元160年的中国和罗马同时发生了瘟疫，很可能是由两国之间的贸易来往造成的。

BUT DISEASES ARE ALSO SPREAD BY WAR, AS ARMIES ENTER PARTS UNKNOWN. (ACCORDING TO ONE THEORY, SYPHILIS CAME TO EUROPE FROM AMERICA WITH THE CONQUISTADORS.)

但是，战争也会传播疾病，因为士兵可能会进入一些不熟悉的地方（有理论声称，梅毒是西班牙征服者从美洲带回欧洲的）。

TODAY, WHEN ANYONE WITH A PLANE TICKET CAN GO ANYWHERE, INCLUDING THE DEEPEST WILDERNESS, CONDITIONS ARE RIPE FOR A NEW PANDEMIC.

今天，一张机票就可以将任何人带到任何地方，包括丛林深处，产生新的流行病的条件已经成熟了。

DISEASE CAN HAVE A MAJOR IMPACT ON POPULATION LEVELS. IN ENGLAND AND FRANCE DURING THE 1200S, FOR EXAMPLE, PLAGUE CARRIED OFF ROUGHLY 1/3 OF THE POPULATION—WHILE THE WARS, REVOLTS, AND FAMINES THAT FOLLOWED IN PLAGUE'S WAKE KILLED ANOTHER 1/3.

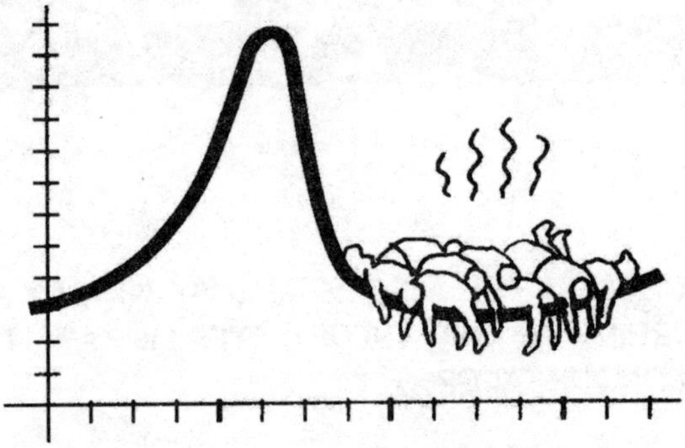

疾病对人口规模有很大的影响。例如，在1200年的英国和法国，瘟疫夺去了大约三分之一人口的生命。接踵而来的战争、骚乱和饥荒又夺去了三分之一人口的生命。

HUMAN HISTORY HAS SEEN THE POPULATION GROWTH CURVE DROP SEVERAL TIMES.

人类历史出现了几次的人口剧增和暴跌。

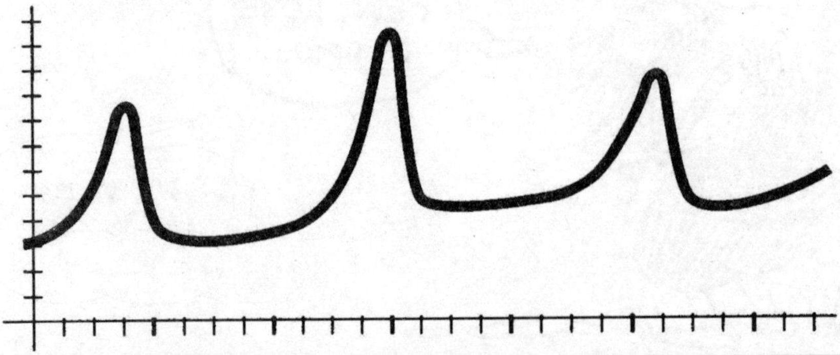

BUT WE REPRODUCE FAIRLY RAPIDLY, AND AT NO TIME WAS THE SURVIVAL OF THE SPECIES SERIOUSLY THREATENED. CERTAINLY, FOR THE PAST 500 YEARS OR SO, THE TREND HAS BEEN IN ONE DIRECTION ONLY:

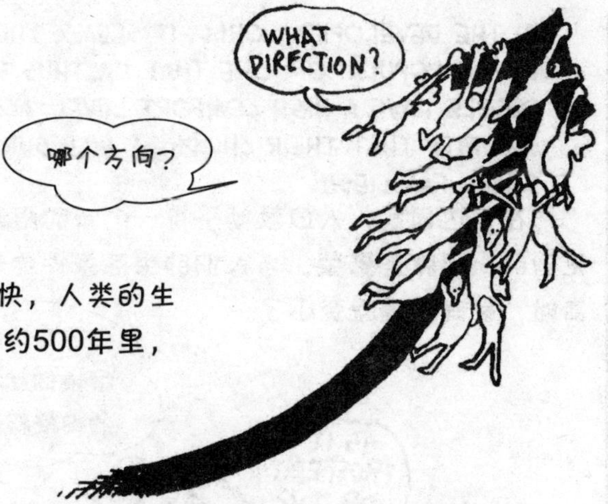

但是，人口数量的重新增长速度也相当快，人类的生存从未受到过严重威胁。当然，在过去的大约500年里，人口增长趋势只朝一个方向变化：

IF YOU LOOK AT POPULATION IN THE WORLD TODAY, YOU GET A SURPRISE: IN THE DEVELOPED WORLD OF THE RICHER COUNTRIES, POPULATION IS BARELY GROWING, WHILE IN THE POORER, DEVELOPING COUNTRIES, IT'S SOARING.

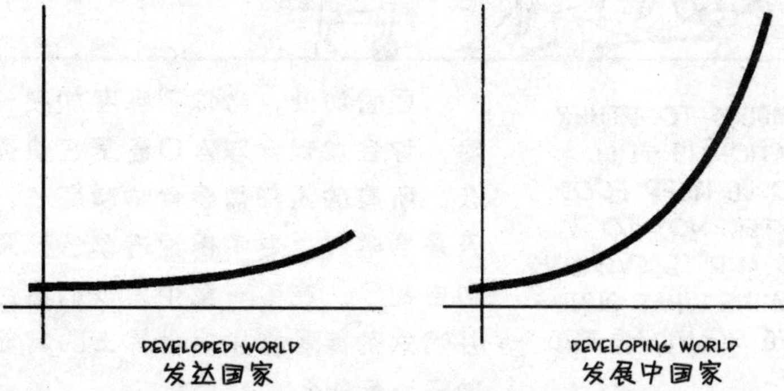

如果你看一下当今世界上的人口，你会大吃一惊：在富裕的发达国家，人口几乎没有增长；而在那些贫穷的发展中国家，人口却在剧增。

IN THE DEVELOPING WORLD, THE LIMITING FACTORS OF WAR, STARVATION, AND DISEASE STILL PLAY A ROLE, BUT THEIR EFFECTS HAVE BEEN MITIGATED BY ANTIBIOTICS AND PUBLIC SANITATION. THE HIGH BIRTH RATE OF THESE MAINLY AGRICULTURAL SOCIETIES THEN PRODUCES HIGH POPULATION GROWTH.

在发展中国家，战争、饥饿和疾病等限制性因素仍然发挥着作用，但抗生素和公共卫生使其影响大大减弱。这些以农业为主的社会人口出生率很高，人口也就急剧增长。

IN THE DEVELOPED WORLD, IT SEEMS, THERE IS A NEW LIMITING FACTOR FOR HUMAN POPULATION, ONE THAT MALTHUS DIDN'T FORESEE: ***PROSPERITY***. WHEN PEOPLE HAVE A HIGH COMFORT LEVEL, ACCESS TO BIRTH CONTROL, AND THE SECURITY THAT THEIR CHILDREN WILL SURVIVE TO ADULTHOOD, THEY HAVE SMALL FAMILIES!!

在发达国家，人口数似乎有一个新的限制因子，这个限制因子是马尔萨斯没有预见到的，那就是**繁荣**。当人们的生活条件优越，并有了节育措施，加上孩子的成活率高时，家庭规模就变小了。

STILL, WHEN YOU ADD THE NUMBERS TOGETHER, YOU FIND THAT GLOBAL POPULATION IS STILL RISING RAPIDLY. ALL THESE PEOPLE NEED FOOD, SHELTER, AND HEAT IN THE WINTER, NOT TO MENTION CARS AND COMPUTERS AND TELEVISIONS. IN THE NEXT CHAPTER, WE EXAMINE WHAT OUR EXPLODING ENERGY NEEDS HAVE MEANT TO THE OTHER SPECIES OF THE WORLD.

尽管如此，当你把数字加在一起时，你会发现全球人口还是在快速增长。所有的人都需要食物和房子，冬天需要暖气，更不用说汽车、计算机和电视了。在下一章中，我们将讨论井喷式的能量需求对世界上的其他物种意为着什么。

CHAPTER 9

BYE BYE, BIOME!

◆第九章◆

再见了，生物群系！

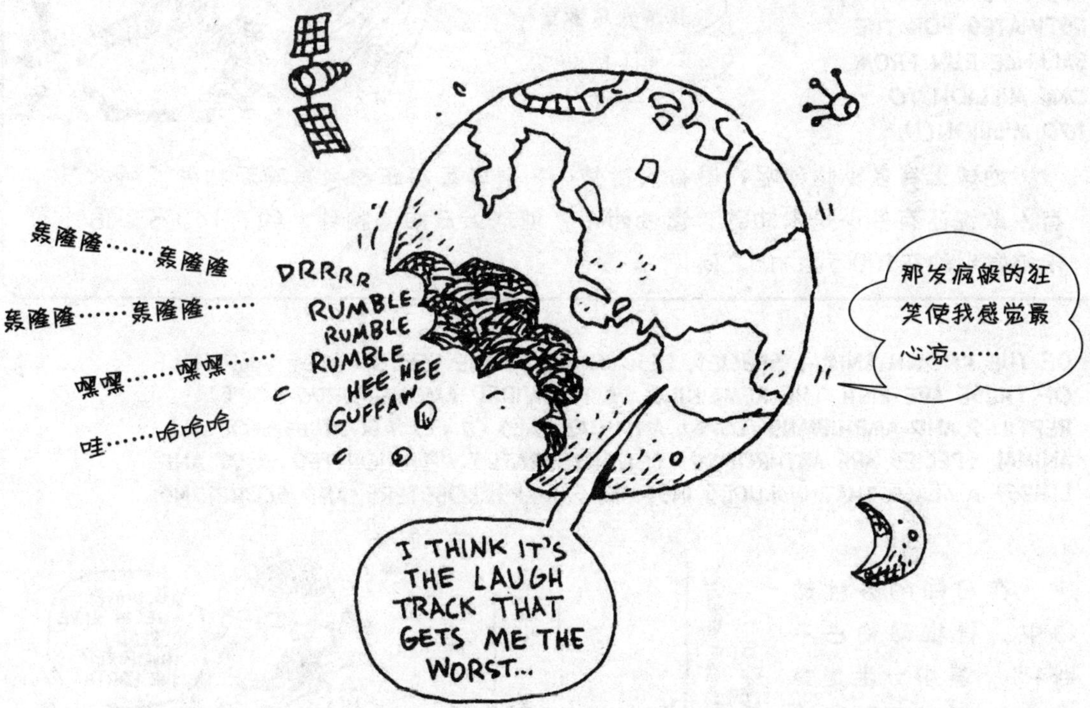

AS THE HUMAN SPECIES HAS DIVERTED BIOLOGICAL RESOURCES TO ITS OWN USE, THE EFFECT ON MOST OTHER SPECIES HAS BEEN DEVASTATING. IN THIS CHAPTER, WE SEE HOW DEFORESTATION, COMMERCIAL HUNTING, AND OTHER HUMAN ACTIVITIES HAVE AFFECTED THE BIOSPHERE... AND WE WARN YOU, IT ISN'T A PRETTY PICTURE!

人类将生物资源据为己有，这对其他物种的影响是毁灭性的。这一章我们将讨论滥伐森林、商业捕猎和其他人类活动是如何影响生物圈的……不过我得事先提醒你，这可不是一幅好景象！

HOW MANY SPECIES ARE THERE ON EARTH? NOBODY KNOWS... EVERY YEAR, SCIENTISTS DISCOVER NEW ONES, AND NO ONE CAN SAY HOW MANY MORE HAVE YET TO BE FOUND. TO DATE, SOME ***1.4 MILLION*** SPECIES HAVE BEEN IDENTIFIED, AND ESTIMATES FOR THE BALANCE RUN FROM ONE MILLION TO ***100 MILLION(!)***.

地球上有多少物种呢？没有人清楚……科学家每年都会发现新物种，所以没有人敢说还有多少是未知的。迄今为止，地球上已知的物种大约有140万，估计未被发现的在100万到1亿之间！

OF THE KNOWN ANIMAL SPECIES, LESS THAN 4% ARE VERTEBRATES, AND HALF OF THESE ARE FISH. THE REMAINING 2% IS DIVIDED AMONG BIRDS (0.8%), REPTILES AND AMPHIBIANS (0.8%), AND MAMMALS (0.4%). ABOUT 85% OF ALL ANIMAL SPECIES ARE ARTHROPODS (INVERTEBRATES WITH JOINTED BODY AND LIMBS), A CLASS THAT INCLUDES INSECTS, SPIDERS, LOBSTERS, AND SCORPIONS.

在已知的各种动物中，脊椎动物占不到4%，其中一半是鱼类。剩下的2%中有0.8%为鸟类，0.8%为爬行动物和两栖动物，0.4%为哺乳动物。大约85%的动物是节肢动物（由体和肢组成的无脊椎动物），包括昆虫、蜘蛛、虾和蝎子。

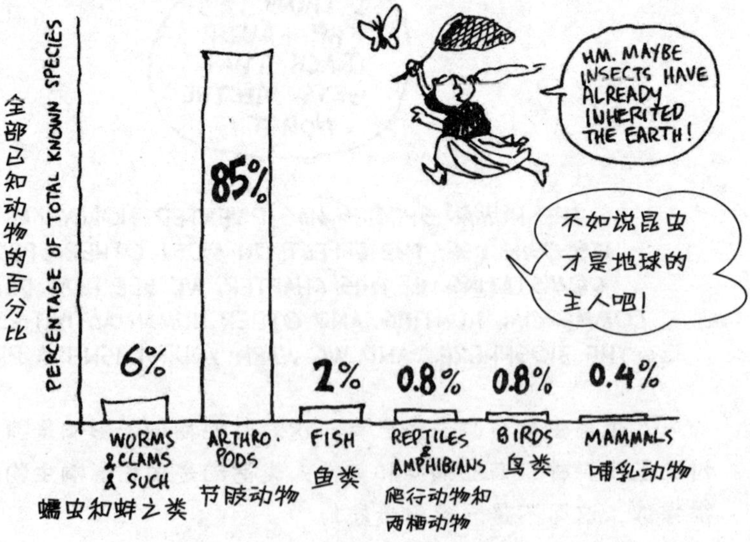

SPECIES ARE NOT SPREAD EVENLY AROUND THE GLOBE. SOME PARTS OF THE WORLD, MOST NOTABLY THE TROPICS, HAVE A MUCH GREATER VARIETY OF FLORA AND FAUNA THAN ELSEWHERE. HERE IS A MAP SHOWING THE REGIONS OF ***MEGADIVERSITY***.

物种在全球的分布是不均匀的。地球上有一些地区，尤其是热带地区，拥有比其他地区更丰富多样的植物区系和动物区系。右边的地图显示了物种**高度多样化**的地区。

ISLANDS IN PARTICULAR MAY BE HIGHLY DIVERSE. MADAGASCAR, FOR EXAMPLE, WHICH WAS ONCE CONNECTED TO THE AFRICAN MAINLAND, BECAME AN ISOLATED HAVEN FOR ANCIENT SPECIES THAT DIED OUT IN OTHER PLACES. TODAY, SOME 10% OF **ALL KNOWN SPECIES** LIVE IN MADAGASCAR, INCLUDING FIVE TIMES AS MANY KINDS OF TREES AS IN ALL NORTH AMERICA!

岛屿上的物种尤其高度多样。例如，曾经与非洲大陆联接的马达加斯加岛，如今成为古代物种的世外桃源，这些物种在其他地区已经灭绝。如今大约**所有已知物种**的10%生长在马达加斯加，这里的树木种类是北美全部树木种类的5倍！

AUSTRALIA SPLIT OFF FROM ASIA WHEN MARSUPIALS WERE COMMON. IN MOST OF THE WORLD THESE PRIMITIVE MAMMALS WERE LARGELY DISPLACED BY PLACENTAL MAMMALS, BUT IN AUSTRALIA THEY FLOURISHED AND DIFFERENTIATED INTO EVERYTHING FROM KANGAROOS TO WOLFLIKE CARNIVORES.

当有袋动物常见的时候，澳洲大陆与亚洲大陆分离。在世界上很多地方，这种原始哺乳动物大部分都被胎盘哺乳动物取代了。但是在澳大利亚，它们却繁衍并分化成很多品种，从袋鼠到与狼相似的食肉动物都有。

*负鼠：产自拉丁美洲的有袋动物。

DESPITE THEIR DIVERSITY, ISLAND ECOSYSTEMS ARE OFTEN FRAGILE. THIS MAKES THEM A DRAMATIC MICROCOSM FOR THE EFFECTS OF HUMAN BEINGS ON THE BIOSPHERE.

岛屿上的物种虽然多样，但很脆弱。因此，人类活动对生物圈的影响，可以在岛屿上表现得淋漓尽致。

ON MADAGASCAR, FOR EXAMPLE, FARMERS HAVE STRIPPED OFF OVER 90% OF THE TROPICAL FOREST, THREATENING THE MOST DIVERSE ECOSYSTEM ON EARTH.

例如，在马达加斯加岛上，农民砍掉了超过90%的热带森林，威胁着地球上最多样的生态系统。

THE SUBSEQUENT EROSION DESTROYS THE QUALITY OF FARMLAND AS WELL. (INDONESIA AND THE PHILIPPINES ARE MOSTLY CLEARED, TOO.)

随之而来的侵蚀破坏了耕地土质（印度尼西亚和菲律宾大部分地区也被开垦了）。

WHEN EUROPEAN SHIPS BEGAN CIRCLING THE GLOBE, THEY WREAKED HAVOC ON MANY AN ISLAND.

当欧洲轮船开始环球航行时，他们对许多岛屿都造成了严重破坏。

ON ISLANDS WITHOUT LARGE PREDATORS, ANIMALS HAD NO URGE TO FLEE THE NEW HUMAN ARRIVALS. THE DODO BIRD, FOR EXAMPLE, WAS SUCH A DOCILE TARGET THAT IT WAS TOTALLY EXTINCT BY 1680.

在没有大型捕食者的岛屿上，当人类到来时，动物并不懂得逃跑。比如说渡渡鸟便是一种非常驯良的猎捕目标，因此到1680年就完全灭绝了。

DITTO THE MOA, THE ELEPHANT BIRD, AND THE TASMANIAN EMU.

同样的情况有恐鸟、隆鸟和塔斯马尼亚鸸鹋的情况也是这样。

EUROPEAN SETTLERS CLEARCUT CARIBBEAN ISLANDS FOR SUGAR PLANTATIONS..

欧洲殖民者为了种植甘蔗而开垦了加勒比群岛。

IMPORTED SHEEP AND CATTLE TO AUSTRALIA, WHERE NONE HAD BEEN BEFORE. NOW THERE ARE 100 MILLION SHEEP AND 8 MILLION COWS IN AUSTRALIA.

绵羊和牛被引进澳大利亚，这里原本是没有牛羊的。如今在澳大利亚共有1亿只绵羊和800万头牛。

AND EVERYWHERE, THEY BROUGHT DOGS, PIGS, AND RATS THAT DEVOURED BIRDS' EGGS AND DISPLACED COMPETITORS.

他们还带来了狗、猪和老鼠，这些动物吃掉了鸟蛋并取代了鸟类的天敌。

MODERN COMMERCIAL "HUNTER-GATHERERS" ASSAULTED SEABIRD POPULATIONS BY HARVESTING EGGS, MEAT, AND FEATHERS FROM THEIR ISLAND NESTING GROUNDS.

现代商业“捕猎—采集者”在岛上的鸟巢里掏鸟蛋，获取鸟肉和羽毛，严重威胁到了海鸟的数量。

AT THE TURN OF THE CENTURY, MILLIONS OF PROCESSED ALBATROSS EGGS WENT INTO U.S. ARMY RATIONS, AND THE ALBATROSS LAYS JUST ONE EGG PER YEAR. THE ALBATROSS HAS SURVIVED, BUT THE GREAT AUK, HUNTED FOR MEAT AND FEATHERS, WENT EXTINCT BY 1844.

在世纪之交，数百万只加工好的信天翁蛋作为美味佳肴分配给美国军队，可这些信天翁每年只产一枚蛋。虽然信天翁幸存下来了，但是人类为了贪得鸟肉和羽毛，残酷捕杀大海雀，到1844年，大海雀已全部灭绝。

BEFORE YOU CONDEMN YOUR OWN SPECIES FOR ITS WANTON BEHAVIOR, NOTE THAT MOST OF THIS DESTRUCTION WAS DONE—AND CONTINUES TO BE DONE—TO PROVIDE PEOPLE WITH FOOD, CLOTHING, SHELTER, AND OTHER NECESSITIES.

在你指责当时人类的这种肆无忌惮的劣行之前，请注意，破坏已经发生，而且会继续发生——目的是给人类提供食物、衣服、住所和其他生活必需品。

IN THE DAYS BEFORE PLASTIC AND PETROLEUM, PEOPLE USED WHALE-OIL LAMPS, FIREWOOD AND CHARCOAL, BEAVER HATS, BADGER-BRISTLE BRUSHES, DEERSKIN GLOVES, LOG HOUSES, AND WALRUS-HIDE BULLETPROOF VESTS.

在塑料和石油被发现之前，人们使用鲸油灯、木柴和炭、海狸帽、獾毛刷、鹿皮手套，住的是木头房子，穿的是海象皮防弹背心。

ON THE OTHER HAND, IT IS A LITTLE HARDER TO JUSTIFY KILLING ELEPHANTS FOR IVORY PIANO KEYS, TORTOISES FOR FORTUNE-TELLING (IN CHINA), CHAMOIS FOR SILVER-POLISHING RAGS, OR OSTRICH, FLAMINGOS, BIRDS OF PARADISE, AND EGRETS JUST FOR THEIR ORNAMENTAL FEATHERS.

另外，人类更没有道理为了象牙琴键而杀死大象，为了算命而杀死乌龟，为了得到擦拭银器的布而杀死臆羚，或者仅仅是为了得到装饰性的羽毛而杀死鸵鸟、火烈鸟、天堂鸟和白鹭。

THE CIVILIZED SYSTEM THAT PRODUCES THIS SLAUGHTER IS

COMMERCIAL HUNTING

(AND FISHING, FOR THAT MATTER).

产生这种屠杀的文明体系是**商业性捕猎**

（从某种意义上说，捕鱼也是如此。）

UNLIKE THE SUBSISTENCE HUNTER, WHO KILLS ANIMALS FOR USE BY HIMSELF AND HIS FAMILY, THE COMMERCIAL HUNTER SERVES A DISTANT ***MARKETPLACE.*** FOR EXAMPLE, THE PLAINS INDIANS KILLED A FEW BISON AND THEN USED MEAT, HIDE, AND ALL, WHEREAS ***BUFFALO BILL*** SLAUGHTERED THE ANIMALS WHOLESALE SO BUTCHERS COULD CUT OFF THE HUMPS AND TONGUES AND SHIP THEM TO RAILROAD WORKERS FOR DINNER, WASTING THE REST.

商业捕猎与生存捕猎不同，生存捕猎只是为了供自己和家庭使用，而商业捕猎则是为了服务更远的**市场**。例如，平原印第安人捕杀少数几头野牛，其肉、皮和剩下的所有东西都被派上用场，然而**野牛比尔**(名为威廉·科迪，美国陆军侦察兵，因擅捕野牛而得名，他曾在8个月内杀死近500头野牛)大规模屠杀野牛，让屠夫只切取隆肉*和舌头，运到铁路工人那儿供他们食用，其余的部分都扔掉了。

*隆肉：野牛肩部中央显著凸起的部位。

GENERALLY SPEAKING, THE SUBSISTENCE HUNTER MARSHALS HIS RESOURCES CAREFULLY. WHEN ONE SOURCE OF FOOD RUNS LOW, HE SWITCHES TO SOMETHING ELSE. THIS STABILIZES THE ECOSYSTEM.

总的来说，生存性捕猎者谨慎地利用资源。当一种食物源减少时，便转向另一种。这样可以稳定生态系统。

COMMERCIAL HUNTING, BY CONTRAST, IS INHERENTLY UNSTABLE. WHEN AN ANIMAL RESOURCE BECOMES SCARCE, ITS PRICE GOES UP, DRIVING THE HUNTER TO HUNT EVEN MORE.

相反，商业捕猎本质上就不稳定。当一种动物资源越稀缺时，其价格就越会上涨，猎人捕猎的欲望就更强，甚至变得更疯狂。

BESIDES, WHAT'S THE COMMERCIAL HUNTER TO DO? HE'S A ***SPECIALIST***... HUNTING IS HIS JOB (AGAIN, UNLIKE THE HUNTER-GATHERER, WHO IS MORE OF A GENERALIST).

此外，想想看商业捕猎者是什么人呢？他是一位**专家**…

捕猎可是他的职业（这也与生存性狩猎—采集者不同，生存捕猎者更像是多面手）。

NOTE: THIS IS NOT TO BE CONFUSED WITH HUNTING FOR SPORT. SPORT HUNTING HAS ALMOST ALWAYS BEEN CLOSELY REGULATED, AND SPORTSMEN LIKE THEODORE ROOSEVELT WERE AMONG THE FIRST CONSERVATIONISTS.

注意：这些不能与狩猎运动混淆。狩猎运动现在几乎都在严格管理之下进行，西奥多·罗斯福这样的狩猎运动员是第一代环境保护主义者的一员。

THE STARKEST EXAMPLES OF COMMERCIAL HUNTING'S EXCESSES COME FROM THE EUROPEAN COLONIZATION OF NORTH AMERICA, WHERE PREVIOUSLY THE INHABITANTS HAD MAINTAINED AN ENVIRONMENT OF VAST FORESTS AND ABUNDANT WILDLIFE.

过度商业捕猎最明显的例子发生在欧洲对北美的殖民统治时期。在这之前，北美的森林密布、野生动物丰富，这里的居民一直保护着这种环境。

THE NUMBERS ARE STAGGERING: 379,012 POUNDS OF SPERMACETI (WHALE GLOP) EXPORTED FROM THE EAST COAST IN ONE YEAR (1770)... BETWEEN 121,355 AND 612,000 DEERSKINS SHIPPED FROM SOUTH CAROLINA EVERY YEAR BETWEEN 1706 AND 1748... 500,000 SNOWY EGRETS FROM VENEZUELA IN 1848... 130,000 BIRD SKINS IN 1892 FROM ONE FLORIDA FEATHER MERCHANT ALONE... 118,000 SEA OTTER PELTS IN 1856... FEATHERS FROM 48,000 CONDORS ON DISPLAY IN LONDON SHOPS IN 1913...

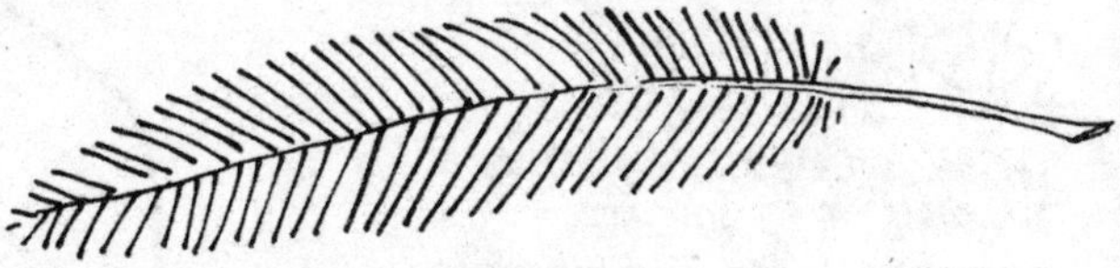

而其中的数字大得让人吃惊：仅在1770年这一年，从东海岸出口的鲸脑油就有379,012磅*……在1706年到1748年之间，每年有121355到612000张鹿皮从南卡罗来纳州运走。1848年，有500000只雪鹭从委内瑞拉运出。单单从一个佛罗里达商人那里就出口了130000张鸟皮……1856年，118000张水獭毛皮……1913年有48000只神鹫的羽毛在伦敦商店里被展示……

*1磅=0.45359237千克，下同。

IN ALL, HUNTERS EXTERMINATED AN ESTIMATED TOTAL OF ***60 MILLION BISON, 200 MILLION BEAVER, 5 BILLION PRAIRIE DOGS,*** AND UNCOUNTED BILLIONS OF PASSENGER PIGEONS.

捕猎者消灭了大约6千万头野牛，2亿只海狸，50万只土拨鼠和几十亿只旅鸽。

TO DESTROY AN ECOSYSTEM, IT IS NOT NECESSARY TO HUNT EVERY SPECIES TO DEATH. REMOVAL OF JUST ***ONE*** SPECIES MAY HAVE A PROFOUND EFFECT ON OTHERS, JUST AS REMOVING THE KEYSTONE HAS AN EFFECT ON AN ARCH.

想要毁灭一个生态系统，并不需要把每个物种都灭绝。只要将一个物种灭绝，就会对其他物种带来巨大的影响，这就如同拆掉拱顶石对拱门的影响一样。

ARCH WITH KEYSTONE
有拱顶石的拱门

...AND WITHOUT
没有拱顶石后

AND SO WE HAVE THE CONCEPT OF A

KEYSTONE SPECIES.

THESE SPECIES ARE NATURE'S ***ENGINEERS***, WHOSE ACTIVITIES ALTER THE ENVIRONMENT IN WAYS THAT CREATE HABITATS FOR OTHER ANIMALS AND HELP DETERMINE THE ECOLOGY OF THE KEYSTONE SPECIES' SURROUNDINGS.

这样我们就有了**关键种**的概念。这些物种是自然的**工程师**，关键种的活动在很多方面改变着环境，它们为其他动物创造栖息地，并决定关键种周边的生态状况。

AN EXAMPLE IS THE ***BEAVER***, WHOSE DAM BUILDING CREATES WETLANDS AND CONVERTS FORESTS TO MEADOWS. THE BEAVER MAKES ITS ENVIRONMENT MOISTER THAN IT WOULD HAVE BEEN, BEAVERLESS.

以**海狸**为例，它们的筑坝行为创造了湿地，将森林转变成草地。海狸使得它的生存环境比没有海狸的地方空气更潮湿。

BUT THE BEAVER'S PELT MADE IRRESISTIBLE HATS... THE BEAVER WAS HUNTED TO NEAR EXTINCTION BY 1840... AND ALTHOUGH THE ANIMALS HAVE BOUNCED BACK SINCE THEN, THEIR CURRENT POPULATION (AROUND 6-12 MILLION) IS ONLY **5%** OF WHAT IT WAS—AND OUR LANDSCAPE HOLDS MUCH LESS WATER THAN IT USED TO.

但是，海狸的毛皮能够制成奢侈的帽子……到1840年为止，海狸已经被人类捕猎到接近灭绝……从那以后，海狸的数量尽管出现反弹，但当前一共只有大约600万到1200万只，仅仅是以前数量的5%。不难想象土壤中的湿度自然比以前低了很多。

OTHER KEYSTONE SPECIES INCLUDE THE ALLIGATOR (IT DIGS WATER-RETAINING HOLES) AND THE PRAIRIE DOG (ITS UNDERGROUND TOWNS PROVIDE HOMES FOR MANY OTHER ANIMALS, AERATE THE SOIL, AND INCREASE ITS ABSORBENCY).

其他关键种还包括短吻鳄（其活动是挖蓄水洞）、草原土拨鼠（它开凿的地下城堡为其他动物提供了栖身地，还使土壤透气，提高了土壤的吸水性）。

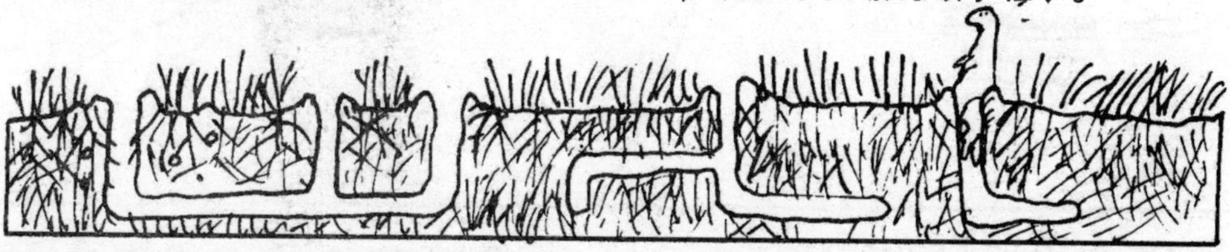

BUT RANCHERS SAW PRAIRIE DOG TOWNS AS NOTHING BUT A BUNCH OF LEG-BREAKING HOLES FOR COWS TO STEP IN. EVEN THOUGH CATTLE ACTUALLY ***PREFERRED*** TO DINE AMONG PRAIRIE DOGS (THE GRASS IS TENDERER THERE), RANCHERS POISONED THE PRAIRIE DOGS TO MAKE THE WORLD SAFE FOR CATTLE.

但在牧场主看来，草原土拨鼠聚集区只不过是一片片绊脚的土洞而已，牛群可能会踏进这些洞中。但牛群实际上更喜欢吃草原土拨鼠区的草（那里的草更水嫩）。牧场主却傻到毒死草原土拨鼠来确保牛群的安全。

IN ADDITION TO OVERHUNTING,

COMPETITIVE EXCLUSION

HAS DEPRIVED WILD SPECIES OF THEIR HABITATS. IN THE U.S., PEOPLE HAVE PLOWED 98% OF THE GRASSLANDS, CUT DOWN 94% OF THE VIRGIN FOREST (ALTHOUGH SOME HAS GROWN BACK), AND FILLED OVER 50% OF THE WETLANDS.

除了过度捕猎外，**竞争排斥**也会使野生物种失去栖息地。在美国，人们已经耕种了98%的草地，砍掉了94%的原始森林（虽然一些已经重新长出来），还填埋了超过50%的湿地。

MIGRATION ROUTES ARE INTERRUPTED BY HIGHWAYS, WHICH BLOCK THE MOVEMENTS OF SMALL ANIMALS AS SURELY AS MOUNTAINS OR RIVERS.

迁徙路线被公路打断，这就像大山或是河流一样阻断了小动物的迁移之路。

EVEN WHERE PATCHES OF HABITAT ARE LEFT, THEY ARE OFTEN TOO SMALL AND DISCONNECTED TO SUSTAIN A VIABLE POPULATION: THE "ISLAND EFFECT."

公路修完之后，为小动物留下一些零星小块栖息地，但是因为面积太小，又不相互联结，因此很难维持一个能自行发展的规模。这就是“岛屿效应”。

TODAY, IN AN EFFORT TO PRESERVE AS MUCH BIODIVERSITY AS POSSIBLE, ECOLOGISTS TRY TO IDENTIFY PLANTS AND ANIMALS AT RISK OF EXTINCTION WHILE THEY'RE STILL ALIVE.

今天，为了尽可能保护生物多样性，生态学家努力识别出那些濒临灭绝但还幸存的植物和动物。

AN ***ENDANGERED SPECIES*** IS ONE WITH SO FEW SURVIVORS THAT THE SPECIES IS AT IMMEDIATE RISK OF EXTINCTION OVER ALL OF ITS RANGE. EXAMPLES INCLUDE THE CALIFORNIA CONDOR (FEWER THAN 100 KNOWN INDIVIDUALS), THE JAVAN TIGER, AND THE AFRICAN WHITE RHINOCEROS.

濒危物种是指幸存者数量太少，以至于全面濒临灭绝的物种。包括加利福尼亚神鹫（不到100只）、爪哇虎和非洲白犀牛。

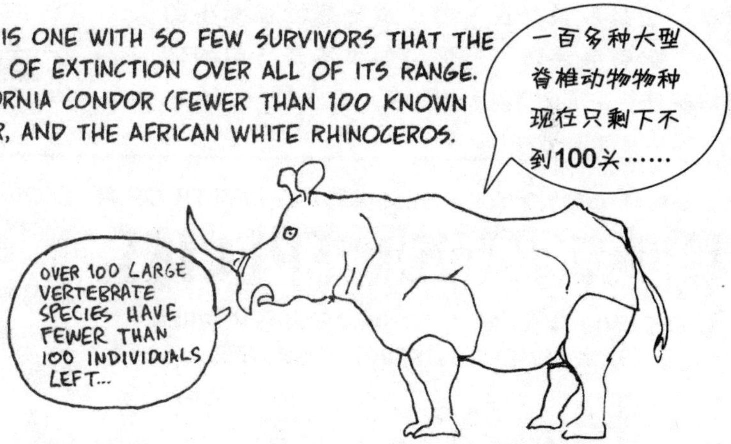

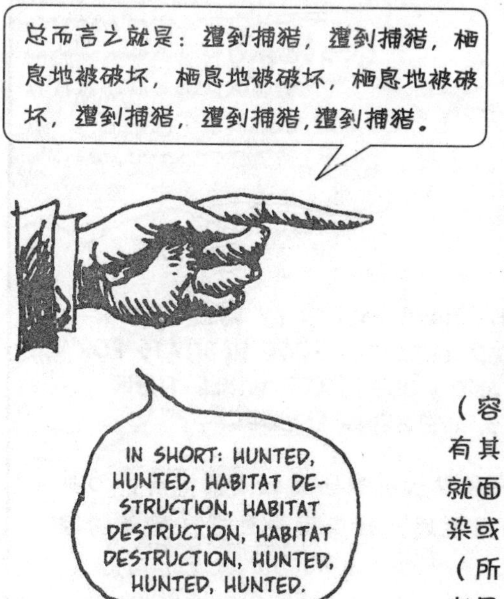

AN ENDANGERED SPECIES MAY HAVE ONE OR MORE OF THESE CHARACTERISTICS: LARGE SIZE, (EASILY HUNTED), FEW OFFSPRING (SLOW TO REPRODUCE), UNIQUE HABITAT (NOWHERE ELSE TO GO), SPECIALIZED DIET (KILL THEIR FOOD AND THEY'RE DEAD), TOP OF THE FOOD CHAIN (POISONED BY POLLUTION, MORE WAYS TO DISRUPT THEIR FOOD SUPPLY), EAT LIVESTOCK (SO THEY'RE HUNTED), MIGRATORY (EASILY HUNTED), OR OTHER RISKY BEHAVIOR (DITTO).

濒危物种可能有以下一个或多个特点：体型较大（容易捕获），后代少（繁殖慢），栖息地独特（没有其他地方可去），食物特别（切断食物来源，它们就面临死亡），在食物链中处于高层（由于环境受污染或其他原因，因而食物供给得不到保障），吃牲畜（所以它们才被捕猎），好迁徙（容易被捕获），或者具有其他危险的行为。

THREATENED SPECIES

ARE THOSE WITH REASONABLE POPULATIONS IN THEIR NATURAL HABITATS, BUT WHOSE NUMBERS ARE FALLING OR HABITATS ARE SHRINKING. EXAMPLES INCLUDE THE BALD EAGLE AND THE GRIZZLY BEAR, WHICH WOULD BOTH BE ENDANGERED, IF NOT FOR CURRRENT ENVIRONMENTAL LAWS.

受威胁物种是那些在自然栖息地中有一定数量、但是数量却在下降、或栖息地在缩小的物种，如白头海雕、灰熊，如果不是当前的环境保护法，它们都将濒临灭绝。

WHEN IT COMES TO ASSESSING THE HEALTH OF AN ECOSYSTEM, WE RELY ON

INDICATOR SPECIES.

THESE ARE THE FIRST TO BE AFFECTED WHEN THEIR ENVIRONMENT IS BEING DEGRADED.

当评价一个生态系统的健康状况时，我们要依靠**指示种**。当环境受到威胁时，它们首先受到影响。

IN THE U.S., MIGRATORY SONGBIRD POPULATIONS HAVE FALLEN BY HALF. THEIR NORTHEASTERN HOMES HAVE BEEN FRAGMENTED, CREATING EDGE HABITATS FOR PREDATORS LIKE COWBIRDS, JAYS, RACCOONS, AND HOUSE CATS, WHILE THEIR WINTER QUARTERS, THE TROPICAL RAINFORESTS, ARE BEING LOGGED.

在美国，迁徙鸣禽的数量已经减半。它们在东北部的家园已经支离破碎，为其捕食者像牛鹂、松鸦、浣熊和家猫等提供了边缘生境。热带雨林是鸟儿冬天的家园，可如今却遭到了砍伐。

AMPHIBIANS ARE ALSO INDICATOR SPECIES, SINCE THEIR THIN SKIN AND PERMEABLE EGGS EXPOSE THEM TO POLLUTANTS IN AIR, WATER, AND SOIL. AMPHIBIAN POPULATIONS HAVE BEEN FALLING WORLDWIDE, EVEN IN HABITATS THAT APPEAR SUPERFICIALLY HEALTHY.

两栖动物也是指示种，它们的薄皮和渗透性强的蛋很容易受空气、水和土壤中污染物的影响。两栖动物的数量在全球范围内正在减少，甚至在那些从表面上看起来环境还算健康的栖息地，其数量也在减少。

IT'S HARD TO ESTIMATE THE RATE OF SPECIES EXTINCTION HISTORICALLY, BUT IT HAS DEFINITELY ACCELERATED IN RECENT YEARS. BY THE 1970S, 1000 SPECIES A YEAR WERE GOING EXTINCT(!), AND BY 1990 THE ANNUAL RATE HAS SOARED TO AN ASTONISHING ***4000-6000***, ACCORDING TO HARVARD BIOLOGIST E.O. WILSON.

我们很难估计历史上物种灭绝的速度，但是在近些年物种灭绝的速度确实上升了。到20世纪70年代为止，每年有1000个物种濒临灭绝！据哈佛大学生物学家E.O.威尔逊的分析，到20世纪90年代，每年濒临灭绝的物种飞涨到4000~6000个。

VIRTUALLY ALL FLORA AND FAUNA ARE DECLINING, AND THOUSANDS OF UNKNOWN SPECIES ARE ALMOST CERTAINLY FALLING WITH THE TROPICAL RAINFORESTS.

几乎所有植物和动物物种的数量都在减少。几乎可以肯定地说，成千上万的未知物种正在随着热带雨林的消失而消失。

IN 1989, ECOLOGISTS AT STANFORD UNIVERSITY CALCULATED AN ESTIMATE OF HOW MUCH OF THE EARTH'S NPP (NET PRIMARY PRODUCTION—I.E., USABLE PLANT BIOMASS) IS NOW USED BY HUMANS.

1989年，斯坦福大学的生态学家通过计算估计出了人类当前使用的净初级生产量（即可利用的植物生物量）。

THEY CONCLUDED THAT WE NOW CONTROL AROUND 39% OF ALL LAND-BASED NPP. WE ACTUALLY EAT ONLY 3%, BUT WE CONSUME THE OTHER 36% IN THE FORM OF CROP WASTES, FOREST CLEARING, DESERTIFICATION, AND SETTLEMENT.

他们得出的结论是，人类控制了大约39%的陆地净初级生产量。但实际上只食用了3%，其余的36%以其他形式被消耗了，如作物废弃部分、森林砍伐、荒漠化、住所建造等。

HUMAN POPULATION IS EXPECTED TO DOUBLE IN THE NEXT 50 YEARS. UNLESS WE CHANGE OUR PATTERNS OF USAGE AND CONSUMPTION, WE WILL BE USING NEARLY ALL LAND-BASED NPP AND MUCH OF THE OCEAN'S, AND WORLD BIODIVERSITY WILL BE A THING OF THE PAST.

人类的人口数量很有可能在以后的50年里增加一倍。如果不改变资源的使用和消费模式，我们将会消耗掉陆地上几乎所有的净初级生产量及海洋净初级生产量的很大一部分，那时候世界上的生物多样性就将成为过去。

WHY SHOULD WE CARE IF GLOBAL DIVERSITY DISAPPEARS? BECAUSE, IN EFFECT, WE ARE PERFORMING AN IMMENSE EXPERIMENT ON THE PLANET, TAKING SYSTEMS APART AND THROWING AWAY THE PIECES. WE KNOW VERY LITTLE ABOUT THE NATURAL WORLD: HOW ANIMALS LEARN, NAVIGATE, OR COMMUNICATE, WHAT MEDICINAL PLANTS STILL LURK IN THE DEPTHS OF THE RAIN FOREST—AND MOST CRITICALLY, WHETHER THE SIMPLIFIED ECOSYSTEM WE ARE SO BUSILY CREATING CAN SUSTAIN ITSELF.

我们为何要担心全球的物种多样性是否会消失？这是因为，实际上我们在地球上正在进行大规模的实验，将系统粗暴地拆开，把它的零件扔得到处都是。我们对自然界的了解并不多。如动物是怎么学习、辨别方向和彼此交流的，什么药草埋藏在热带雨林的深处等。最重要的是，我们忙着建造的这个简单的生态系统是否可以持续。

~ CHAPTER 10 ~

ENERGY WEBS

◆第十章◆

能源网

HUMAN BEINGS, ALONE AMONG ANIMALS, HAVE FOUND WAYS TO USE ***NON-FOOD ENERGY*** IN OUR LIVES. BURNING WOOD AND OIL HEAT OUR HOMES; ELECTRICITY POWERS OUR ELECTRIC TOOTHBRUSHES, LIGHTS OUR BULBS, AND RUNS OUR COMPUTERS; GASOLINE FIRES OUR CARS. WE SQUANDER THE STUFF, ESPECIALLY PETROLEUM PRODUCTS, AS IF THERE WERE NO TOMORROW.

在所有动物中，只有人类发明了在生活中使用**非食物能量**的方法。燃烧木材和石油可以使房间变暖；电能够驱动电动牙刷、使灯泡发光、使电脑运转；汽油能够发动汽车。我们大量消耗这些资源，尤其是石油产品，好像明天就是最后一天。

WHAT IS ENERGY, ANYWAY? (WE'VE MANAGED TO GO 9 CHAPTERS WITHOUT SAYING EXACTLY.) FOR THE MOMENT, LET'S PRETEND THERE ARE ONLY TWO KINDS OF ENERGY, OR AT LEAST TWO WAYS TO THINK ABOUT ENERGY: AS ***HEAT*** AND ***WORK***.

到底什么是能量呢？（我们讲了九章了，但一直都没有确切地说明。）现在，我们暂时假定世界上只有两种能量，或者说两种考虑能量的方式，即**热**和**功**。

ON THE ONE HAND, ENERGY CAN BE THOUGHT OF AS AN AMOUNT OF ***HEAT***. IF YOU BURN FUEL, THE ENERGY RELEASED IS (NEARLY) THE SAME AS THE TOTAL AMOUNT OF HEAT GIVEN OFF. (SOME ENERGY IS ALSO EMITTED AS LIGHT.)

一方面，能量可以被看成是一定量的**热**。如果你燃烧燃料，释放的能量近似于发出的热的总量（一部分能量也会以光的形式被释放）。

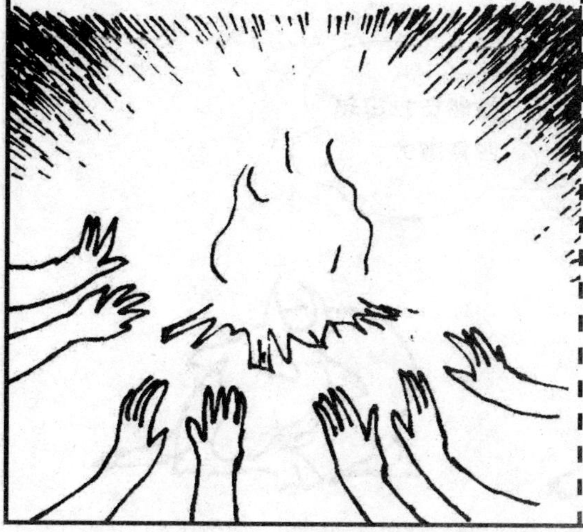

ON THE OTHER HAND, IN MECHANICAL TERMS, ENERGY IS THE SAME AS ***WORK***. IF YOU PUSH SOMETHING WITH A FORCE OVER A DISTANCE, THE ENERGY EXERTED IS (NEARLY) THE PRODUCT OF THE FORCE TIMES THE DISTANCE. (SOME ENERGY IS ALSO DISSIPATED AS HEAT DUE TO FRICTION.)

另一方面，用机械术语讲，能量等同于**功**。如果你用力将一个东西推一段距离，施加的能量近似于力与距离的乘积（由于存在摩擦力，一部分能量便以热量的形式消散掉了）。

Energy = Force × distance

能量 = 力 × 距离

FOR MOST OF HUMAN HISTORY, THESE WERE THE KINDS OF ENERGY PEOPLE NEEDED: HEAT AND WORK.

在人类历史长河的绝大部分时间里，人们所需的能量正是这两种：热和功。

AT FIRST, THERE WAS BASICALLY ONLY ONE SOURCE OF HEAT, AND THAT WAS **BIOMASS**: WOOD, STRAW, OR, WHERE WOOD WAS SCARCE, DRIED COW DUNG (AS IN INDIA), ETC.

起初，世界上基本上只有一种热源，那就是**生物质**，即木材和稻草，在木材短缺的地方（如印度），还使用干牛粪等。

MECHANICAL ENERGY ALSO CAME FROM BIOMASS: ANIMAL POWER--OR BRUTE FORCE--AND THE BRUTE IN QUESTION USUALLY HAD TWO LEGS.

机械能同样来自于生物质，如动物力量或者叫兽力，相关的野兽通常只有两条腿。

EVEN WHEN ANIMALS WERE HARNESSED, THEIR FUEL EFFICIENCY WAS LOWER THAN HUMANS': I.E., THEY CONSUMED MORE FOOD TO DO THE SAME WORK. ECONOMICS DICTATED THAT HUMANS DO MOST OF THE WORK, SAVING ANIMALS FOR THE REALLY HEAVY LIFTING AND HAULING.

即使动物被利用上了，其能效也低于人类。也就是说，做同样多的工作，动物比人吃得更多。根据经济学原理，人类应该做大部分工作，让动物做那些真正的重活，如搬运重物。

EVEN WITH THESE LIMITATIONS, PEOPLE BUILT GREAT CIVILIZATIONS. IN MANY CASES, THE CONSTRUCTION WORKERS WERE SLAVES: HUMAN ENERGY USED EN MASSE AND FED AS LITTLE AS POSSIBLE.

尽管有这些限制，人类还是建立了伟大的文明。在很多情况下，建筑工人就是奴隶：做最多的工作，吃最少的食物。

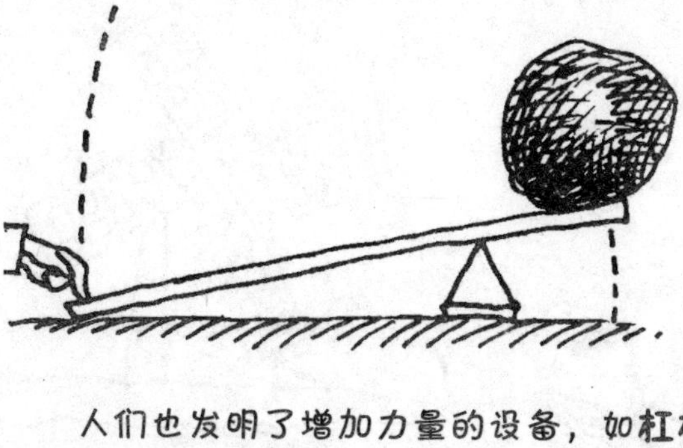

PEOPLE ALSO CREATED STRENGTH-ENHANCING INVENTIONS LIKE THE ***LEVER:*** USING A LEVER, A GENTLE PUSH EXERTED OVER A LONG DISTANCE IS CONVERTED INTO A BIG PUSH EXERTED OVER A SHORT DISTANCE.

人们也发明了增加力量的设备，如**杠杆**。使用杠杆时，在较远处轻轻一推，就能转化为很大的力量。

A FEW TOOLS THAT USE THE PRINCIPLE OF THE LEVER:

应用杠杆原理所制造的工具有：

*曲柄：在连架杆中，能绕其轴线回转360°的物件。

HEAT-ENHANCERS WERE FOUND TOO: WOOD WAS REDUCED TO CHARCOAL, WHICH BURNS HOTTER, AND OVENS, KILNS, AND SMELTERS WERE DESIGNED TO BAKE ANYTHING FROM BREAD TO BRICK TO METAL. WHENEVER POSSIBLE, THESE FACTORIES WERE BUILT NEAR FORESTS, THE FUEL SUPPLY.

增热器也得以发明。将木材制作成木炭后，燃烧时可以产生更高的热量。我们设计了炉子、烧窑、熔炉，用来烤面包、烧砖、炼金属。只要有可能，这些工厂都会建在森林附近，因为森林是燃料供应地。

(AND LET'S NOT FORGET THE MACHINERY DRIVEN BY WATER WHEELS AND WIND POWER.)

（我们也别忘了水轮和风力驱动的机器。）

THAT'S HOW EVERYTHING RAN UNTIL AROUND 1700, WHEN A NEW INVENTION REVOLUTIONIZED TECHNOLOGY: THE

一切都是这样运转的，直到1700年左右，一项新发明带来了技术的大变革，这项发明就是**蒸汽机**。

STEAM ENGINE!

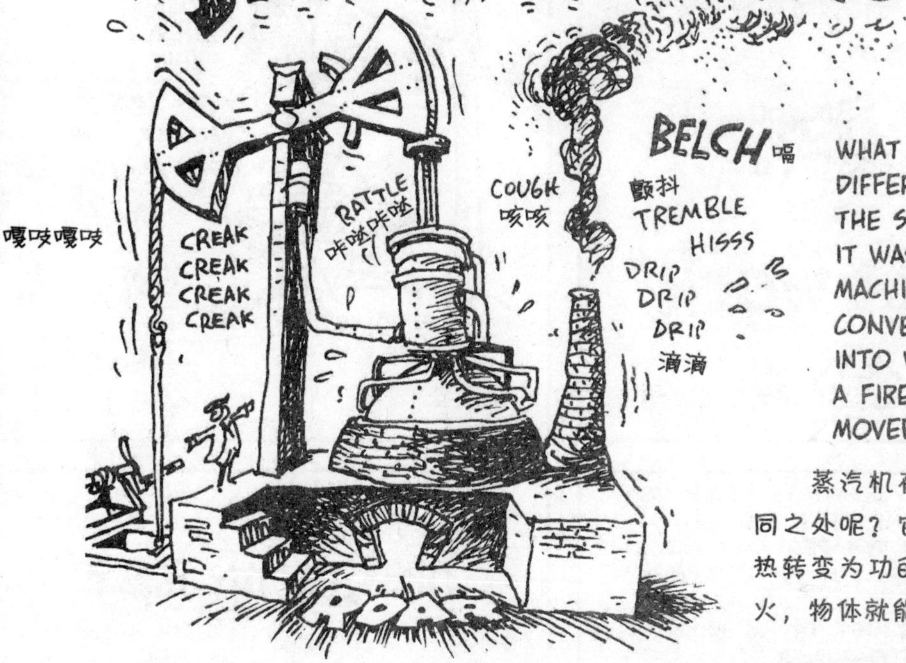

WHAT WAS SO DIFFERENT ABOUT THE STEAM ENGINE? IT WAS THE FIRST MACHINE THAT CONVERTED HEAT INTO WORK! LIGHT A FIRE, AND THINGS MOVED!!!

蒸汽机有什么与众不同之处呢？它是第一个将热转变为功的机器！点上火，物体就能移动！！！

THE STEAM ENGINE HAD TWO GREAT EFFECTS, ONE INTELLECTUAL AND SCIENTIFIC, THE OTHER PRACTICAL.

蒸汽机给人类带来了两大影响：一是智能和科学方面的影响；二是在实际应用方面的影响。

ON THE SCIENTIFIC SIDE, SADI *CARNOT* (1796-1832) WAS INSPIRED TO INVENT THE ***THEORY OF HEAT,*** WHICH TURNED EVENTUALLY INTO THE SCIENCE OF

THERMODYNAMICS.

在科学层面上，萨迪·卡诺（1796—1832）受到启发，创造了热理论，该原理发展成为**热力学**。

THERMODYNAMICS UNITES THE TWO VIEWS OF ENERGY: HEAT AND WORK ARE JUST ***TWO FORMS OF THE SAME THING,*** AND THEY WERE ***INTERCONVERTIBLE.*** YOU CAN CHANGE HEAT INTO MOTION, AND VICE VERSA.

热力学结合了关于能量的两种观点，认为热和功是**同一现象的两种形式**，可以**相互转化**。可以把热转变为运动，反之亦然。

THE STEAM ENGINE IS A ***CONVERTOR:*** ENERGY GOES IN AS HEAT AND IS CONVERTED TO MECHANICAL (OR KINETIC) ENERGY.

蒸汽机是一个**转换器**：能量以热的形式进入，然后转化为机械能（或者动能）。

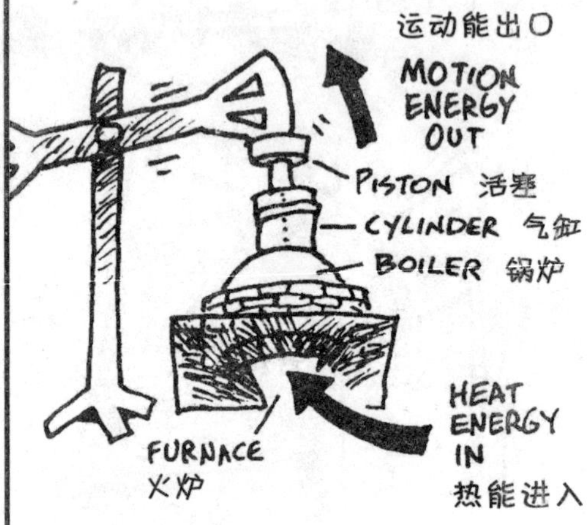

BUT THE CONVERSION IS NEVER 100% EFFICIENT. SOME OF THE BOILER'S HEAT ALWAYS ESCAPES, AND THE MECHANICAL OUTPUT IS ALWAYS SLOWED BY FRICTION, WHICH PRODUCES HEAT. (JUST TRY RUBBING YOUR HANDS TOGETHER.)

但是转化率并不是100%。锅炉里的热量总会渗出一些，摩擦也会降低机械能的输出速度，同时也会产生热量（试着摩擦自己的双手）。

TO SUMMARIZE,

总结一下吧：

heat energy input

⇩

mechanical energy output

+

waste heat

热能输入

⇩

机械能输出

+

废热

IT WAS SOON RECOGNIZED THAT ALL ENERGY, NOT JUST HEAT AND WORK, WERE REALLY THE SAME. GRAVITY, ELECTRICITY, CHEMICAL REACTIONS, SOUND, LIGHT: ALL WERE TYPES OF ENERGY THAT MIGHT BE CONVERTED, CHAMELEONLIKE, ONE TO THE OTHER.

克劳修斯*

很快人们便发现所有的能量都是一样的，不仅仅是热和功。所有的能量，如重力、电力、化学反应、声音和光等都是可转化的能源形式，就像变色龙一样，它们可以从一种形式转化为另一种形式。

FOR EXAMPLE, ***GRAVITATIONAL ENERGY*** MAKES WATER FLOW DOWNSTREAM, TURNING A WATER WHEEL, WHICH CAN DO WORK, WHETHER GRINDING GRAIN OR HOISTING MINERALS FROM A MINE.

例如，**重力能**使水向下流，使水轮转动、工作，或者碾碎谷物，或者将矿物质从矿山中吊起。

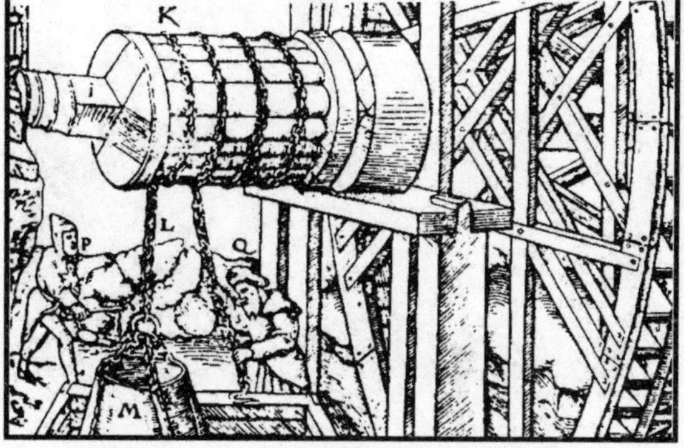

AGAIN, THE MILL IS A ***CONVERTOR,*** CHANGING GRAVITATIONAL ENERGY INTO MECHANICAL ENERGY.

同样，磨坊也是**转换器**，能将重力能转化为机械能。

ALSO, BY THE ***SECOND LAW OF THERMODYNAMICS*** (SEE P. 86), EVERY MACHINE MUST ***WASTE SOME ENERGY,*** MAINLY DISSIPATING IT INTO THE ENVIRONMENT AS HEAT.

同时，根据**热力学第二定律**（见第86页），每一台机器都会**消耗一些能量**，主要是以热量的形式消散到环境中。

BOLTZMANN

玻尔兹曼*

IN OTHER WORDS, ALL ENERGY TECHNOLOGY CONSISTS OF ***CONVERTORS,*** WHICH ARE ALWAYS LESS THAN 100% EFFICIENT.

换句话说，所有的能源技术都包括**转换器**，总是无法达到100%的效率。

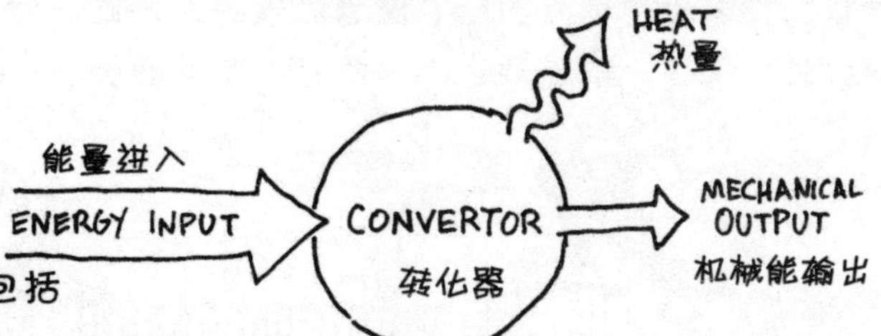

*克劳修斯：德国物理学家，气体理论和物理学的主要奠基人。
*玻尔兹曼：奥地利物理学家，热力学和统计物理学的奠基人之一。

THE ANSWER TURNED OUT TO BE THIS UNIMPRESSIVE-LOOKING STUFF:

到最后，答案却是这种难看的东西：

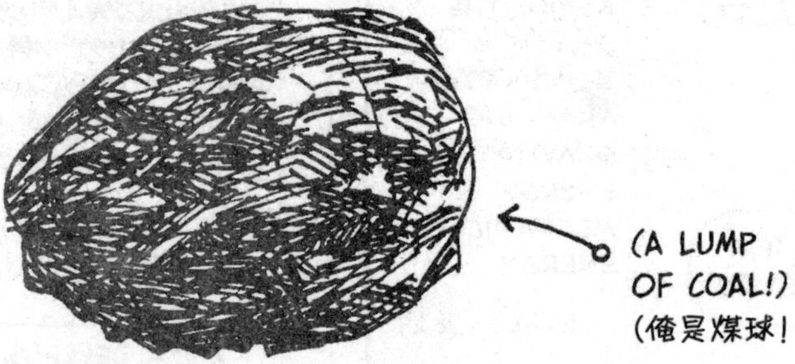

THE STEAM ENGINE WAS FOLLOWED BY AN EXPLOSION, AND NOT ONLY OF BOILERS... BUT ALSO AN EXPLOSION OF INVENTIONS AND DISCOVERIES.

随蒸汽机而来的是一场大爆炸，不仅仅是锅炉大爆炸，而且是发明和发现大爆炸。

OVER THE NEXT 200 YEARS CAME A SERIES OF NEW CONVERTORS, NEW TECHNOLOGIES, NEW FUELS, EVEN NEW FORMS OF ENERGY, LIKE ELECTRICITY.

在随后的200年里，出现了一系列的新转换器、新科技、新燃料，甚至是新的能源形式，如电力。

JUST LIKE THE INVENTION OF AGRICULTURE, THE USE OF FOSSIL FUELS PUT NEW ENERGY IN HUMAN HANDS. THE EFFECTS WERE SIMILAR: ***MORE PEOPLE*** AND ***MORE ORGANIZATION.*** THIS TIME IT WAS CALLED THE

INDUSTRIAL REVOLUTION.

正如农业的发明一样，化石燃料的使用带给人类新的能源。影响是相似的，即**更多的人口和更多的组织**。这次影响被称为**工业革命**。

AN ENERGY NETWORK RESEMBLES A FOOD WEB. START WITH AN ENERGY SOURCE, LIKE AN OIL WELL... THE FUEL MUST BE EXTRACTED, USING UP SOME ENERGY IN THE PROCESS (TO RUN THE PUMP, FOR EXAMPLE)... TRANSPORTED, AGAIN AT SOME ENERGY COST... STORED... DISTRIBUTED... AND CONVERTED TO USEFUL FORM. EACH STAGE WASTES SOME ENERGY AND LOWERS THE OVERALL EFFICIENCY OF THE SYSTEM.

能源网与食物网相似。能源网始于一种能源，如油田……燃料必须首先被提取出来，在提取过程中要消耗一定能量（例如，油泵的运转）……然后将油运输出去，运输过程又要消耗一定能量。石油在储存、分销、转化为有用的形式等过程中也要消耗一定能量。由于每个阶段都要消耗一定能量，整个系统的效率便降低了。

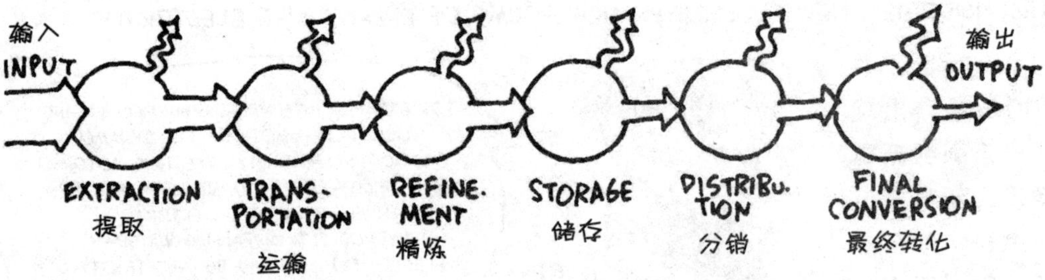

JUST AS WITH FOOD WEBS, THE EFFICIENCY AT EACH STAGE IS THE PERCENTAGE OF ENERGY THAT SURVIVES THE PROCESS IN USABLE FORM. FOR EXAMPLE, TO COMPUTE THE EFFICIENCY OF MOVING A TANKER FULL OF OIL FROM ONE PLACE TO ANOTHER, YOU HAVE TO ACCOUNT FOR THE OIL USED IN DRIVING THE TRUCK.

与食物网一样，每一个阶段的效率是指完成整个过程之后剩下的可用能量的百分比。例如，计算将一辆装满油的油罐车从一个地方运输到另一个地方的效率，把你必须驱动卡车消耗的油量计算进去。

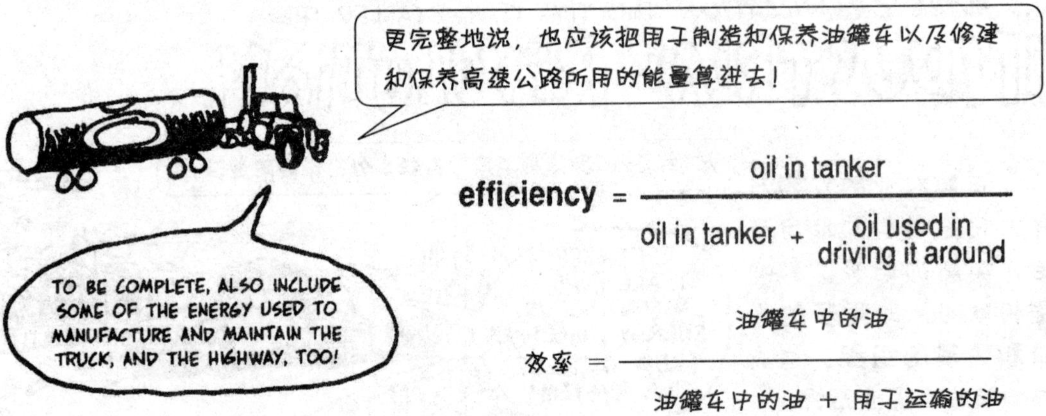

$$\text{efficiency} = \frac{\text{oil in tanker}}{\text{oil in tanker} + \text{oil used in driving it around}}$$

$$\text{效率} = \frac{\text{油罐车中的油}}{\text{油罐车中的油} + \text{用于运输的油}}$$

THE OVERALL EFFICIENCY AT ANY STAGE OF THE ENERGY NETWORK IS THE PRODUCT OF THE EFFICIENCIES OF ALL THE PREVIOUS STAGES. HERE IS A HYPOTHETICAL EXAMPLE:

能源网中任何一个阶段的总效率是以前各个阶段效率的乘积。假定这样这一个例子：

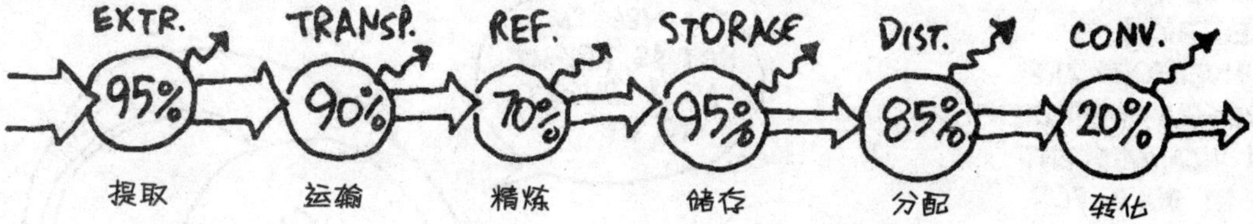

$$0.95 \times 0.9 \times 0.7 \times 0.95 \times 0.85 \times 0.2 = 0.097 \sim \mathbf{10\%}$$

IF YOU THINK THIS LOOKS INCREDIBLY WASTEFUL, YOU ARE RIGHT! A TYPICAL AUTOMOBILE ENGINE IS AT MOST ABOUT 20% EFFICIENT AS A CONVERTOR. IF THE SYSTEM THAT EXTRACTS, REFINES, AND DISTRIBUTES PETROLEUM PRODUCTS IS 50% EFFICIENT (A REASONABLE ESTIMATE), THEN THE ENERGY ACTUALLY DELIVERED TO A CAR'S WHEELS IS NO MORE THAN *10%* OF THE CHEMICAL ENERGY IN THE ORIGINAL PETROLEUM.

你认为这样极其浪费吧，没错！一个标准的汽车发动机作为一个转换器，其效率最多只有20%。如果提取、精炼和分销石油产品的系统效率是50%（一个合理的估计），那么，传输给汽车轮子的能量实际上不超过最初石油中化学能的10%。

*1加仑=4.54609升，下同。

BUT WHEN IT COMES TO INEFFICIENCY, THE KING OF CONSUMPTION, THE SULTAN OF SLACK, THE WAZIR OF WASTE, IS **ELECTRICITY.**

说到低效率、耗能王、马虎大王、浪费之最，非**电力**莫属。

ONE REASON IS SIMPLE: MANY ELECTRICAL CONVERTORS ARE GROSSLY INEFFICIENT. AN INCANDESCENT LIGHT BULB, FOR EXAMPLE, HAS AN EFFICIENCY OF AROUND 5%. ***95% OF THE ELECTRIC ENERGY USED BY A LIGHT BULB IS WASTED AS HEAT.***

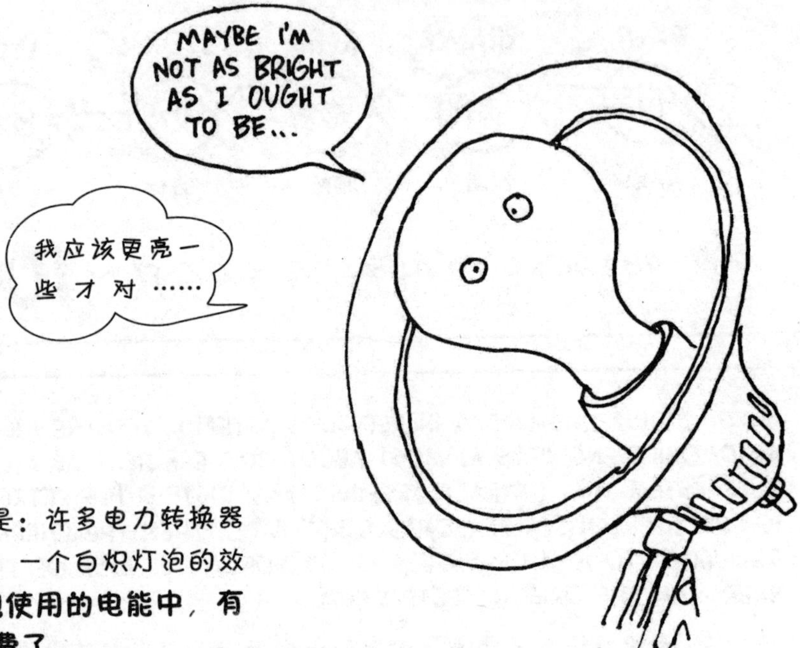

一个简单的原因就是：许多电力转换器总体上效率不高。例如，一个白炽灯泡的效率大约是5%。**一个灯泡使用的电能中，有95%是以热量的形式浪费了。**

THAT 5% COMES AFTER ALL THE LOSSES IN EXTRACTION, TRANSPORT, ETC., SO AN ELECTRIC LIGHT MAY ACTUALLY USE ***LESS THAN 2%*** OF THE ORIGINAL FUEL'S ENERGY CONTENT.

这5%的效率还要减去提取和运输过程中的损耗。因此，一个电灯泡可能实际上使用了**不到**原始燃料能量的2%。

THIS IS TRUE ONLY OF INCANDESCENT BULBS. FLUORESCENT LIGHTS ARE SOMEWHAT MORE EFFICIENT, AND SOME EXPERIMENTAL SULFUR-BASED LAMPS APPROACH 100% EFFICIENCY.

这种情况只适用于白炽灯。荧光灯相对来说效率会高一些，尚处于实验阶段的微波硫灯效率接近100%。

ELECTRIC POWER ALSO WASTES ENERGY BY INSERTING ANOTHER LEVEL OF CONVERSION IN THE ENERGY NETWORK. INSTEAD OF CONVERTING FOSSIL FUEL DIRECTLY TO USEFUL FORM (MOTION, HEAT, LIGHT, ETC.), AN ELECTRIC POWER PLANT FIRST CONVERTS FUEL TO ELECTRICITY, WHICH THEN FLOWS THROUGH WIRES (LOSING SOME ENERGY IN THE PROCESS!) TO ELECTRICAL APPLIANCES, THE ULTIMATE CONVERTORS FOR USE.

电力也会损耗能量，因为在能源网中要多一个过程。电能不像化石能源那样将化石燃料直接转变为有用的形式（运动、热量、光等），发电站首先将燃料转变成电力，电力要通过电线传输给电器（这个过程会损失一定能量！），即最终的转换器。

TO ITS CREDIT, ELECTRICITY IS FLEXIBLE, VERSATILE, EASILY DISTRIBUTED, AND CLEAN—AT LEAST, AT THE CONSUMER LEVEL, IT'S CLEAN. IT CAN ALSO BE GENERATED BY WATER OR WIND POWER, REDUCING OUR DEPENDENCE ON FOSSIL FUELS, AND FOR SOME USES, IT'S HARD TO IMAGINE AN ALTERNATIVE.

电力也有功劳，它很灵活、多样、容易传输，还很清洁——至少到达消费者层面时很干净。而且，水和风也可以用来发电，这样就降低了我们对化石燃料的依赖，电有很多用途，很难想象用别的能源来替代它。

FOR MOST OF THE CENTURY BETWEEN 1870 AND 1970, ENERGY SUPPLIES HAVE BEEN RELATIVELY CHEAP AND ABUNDANT, ESPECIALLY IN THE UNITED STATES, WHICH HAD VAST RESERVES OF OIL AND COAL.

1870~1970年的大部分时间中，能量供给相对便宜，而且很充足，尤其是在美国，石油和煤的储存量特别丰富。

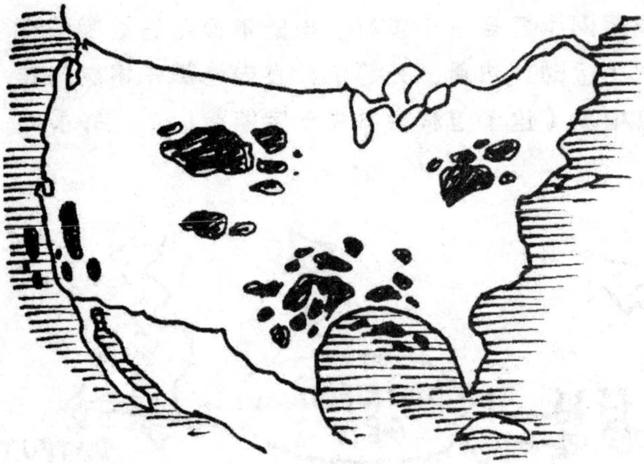

UNDER THESE CONDITIONS, THE U.S.A.'S NATIONAL ENERGY STRATEGY WAS SIMPLE: EXTRACT IT FAST, USE LOTS OF IT, AND DON'T WORRY TOO MUCH ABOUT EFFICIENCY OR POLLUTION.

在这种条件下，美国的国家能源策略很简单：快速提取、大量使用，不用过多地担心利用效率和污染。

(WE HAVEN'T DISCUSSED POLLUTION YET, BUT MOST CONVERTORS PRODUCE NOXIOUS GASES, ASH, OR OTHER COMBUSTION BY-PRODUCTS. MORE ON THIS LATER!)

（我们还没有讨论过污染，但是大多数的转换器都会产生有毒气体、灰尘或者其他由燃烧产生的物质。随后会详细介绍！）

WORLD ENERGY CONSUMPTION ROSE *SIXTY-FOLD* BETWEEN 1860 AND 1985—ESPECIALLY IN THE U.S.A, WHERE CHEAP GAS CREATED A CAR CULTURE...

1860~1985年，世界能源消费量上升了60倍，这在美国尤为典型，便宜的汽油创造了汽车文化……

AND CHEAP ELECTRICITY PRODUCED **ALUMINUM,** WHICH CONSUMES **15 TIMES AS MUCH ENERGY** TO MANUFACTURE AS STEEL. A **SINGLE ALUMINUM BEVERAGE CAN** CONSUMES THE SAME ENERGY AS A 100-WATT LIGHT BULB BURNING FOR FOUR HOURS... AND WE PRODUCE **100 BILLION ALUMINUM CANS A YEAR!!**

电价低，可便于大量生产**铝**，生产**铝**所需要的**能量**是生产钢的15**倍**。生产**一个铝制易拉罐**所消耗的能量相当于一个100瓦的电灯泡4个小时使用的能量，而我们**一年就生产**1000**亿个铝制易拉罐**！！

WESTERN EUROPE HAD LESS OIL AND COAL THAN NORTH AMERICA, BUT COUNTRIES LIKE BRITAIN, FRANCE, AND GERMANY HAD THE FINANCIAL AND MILITARY POWER TO GAIN ACCESS TO OVERSEAS ENERGY RESERVES.

西欧比北美的油和煤储藏量少，但英国、法国和德国等国家有足够的经济和军事实力去获取海外的能源。

WITH ITS MORE EXPENSIVE OIL, EUROPE HAS MANAGED TO BUILD A HIGHLY INDUSTRIAL ECONOMY WITH AROUND HALF THE PER-CAPITA ENERGY CONSUMPTION OF THE UNITED STATES.

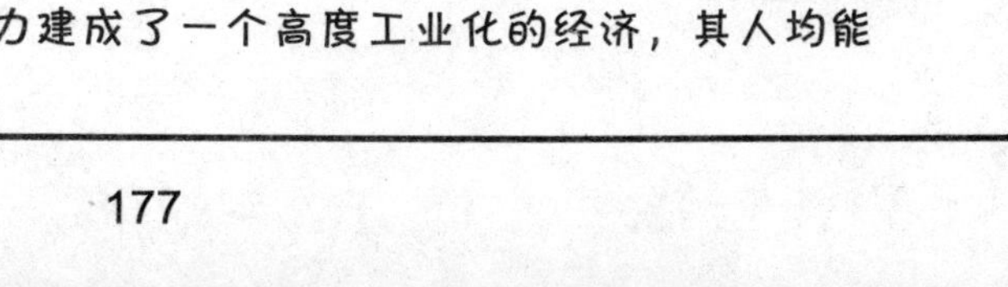

虽然油价更高，但欧洲还是努力建成了一个高度工业化的经济，其人均能源消费量大概是美国的一半。

AS THE WORLD EMBRACES THE JOYS OF INDUSTRIALIZATION, THE DEMAND FOR FOSSIL FUELS HAS RISEN RAPIDLY.

世界各国无不欣然接受工业化所带来的欢乐。因此，对化石燃料的需求陡然上升。

IF WE CONTINUE TO CONSUME OIL AT THIS RATE, IT WILL ALL BE GONE BY THE YEAR 2030. COAL IS GOOD FOR PERHAPS ANOTHER CENTURY AFTER THAT.

如果我们继续以这种速度消费，到2030年石油就会枯竭。此后，煤可能还可以维持一个世纪。

WOULD YOU CALL THIS AN

energy crisis?

WHY NOT?

你会说这是一种**能源危机**吗？不是吗？

FINDING SOLUTIONS TO THE ENERGY PROBLEM IS COMPLICATED BY THE FACT THAT ENERGY CONSUMPTION IS SO UNEQUAL. THE UNITED STATES, WITH 4% OF THE WORLD'S POPULATION, USES ABOUT **25%** OF ALL THE ENERGY. IN FACT, ONE ESTIMATE SAYS WE USE 10% OF THE GLOBAL ENERGY BUDGET JUST DRIVING TO WORK AND BACK. OBVIOUSLY, WE HAVE SOME WASTE TO CUT!

想要找到解决能源问题的方法很复杂，因为能源消耗很不均衡。美国人口只占世界人口的4%，但使用的能源却是世界总能源的25%。实际上，根据估算，仅开车上班就使用了全球能源预算的10%。很明显，我们必须削减消耗量。

BUT IN MUCH OF THE WORLD, PEOPLE ARE POOR AND HAVE LITTLE ACCESS TO ENERGY. THE AVERAGE CITIZEN OF INDIA, FOR EXAMPLE, USES AROUND **2-3%** THE ENERGY CONSUMED BY THE AVERAGE AMERICAN. FOR THE 2 BILLION PEOPLE WHO DEPEND ON FIREWOOD FOR FUEL, THE ENERGY CRISIS LOOKS ENTIRELY DIFFERENT. THEY WANT MORE HIGH-QUALITY ENERGY.

但是在世界上很多地方，人们还很贫穷，几乎买不起能源。例如，一个普通印度人的能源消费量大约是一个普通美国人能源消费量的2%～3%。对于世界上20亿以木柴作为燃料的人来说，能量危机完全是一幅不同的景象了。他们需要的是质量更高的能源。

WHATEVER THE OVERALL SOLUTION TO THE ENERGY CRISIS MAY BE, IT HAS TO INCLUDE STEPS BY THE INDUSTRIALIZED, HIGH-CONSUMING NATIONS TO REDUCE THEIR FOSSIL FUEL USAGE. WHERE CAN THE SAVINGS COME FROM: LET'S SIMPLIFY THE DIAGRAM OF AN ENERGY SYSTEM, LUMPING ALL THE CONVERSION AND TRANSPORTATION STEPS INTO ONE:

不管最终解决能源危机的整体方案是什么，高消费工业化国家必须采取措施，降低化石燃料的使用量。如何才能节约能源呢？让我们将能源系统简单地用左图表示，将所有的转变和运输步骤合并为一个：

THERE ARE ONLY THREE POSSIBLE AREAS OF POTENTIAL SAVINGS, AREN'T THERE? THERE'S THE INPUT, THE WASTE, AND THE OUTPUT!

节能只能从三个方面入手，不是吗？即输入、废料和输出！

output:

THIS INVOLVES MAKING CHOICES THAT USE LESS ENERGY: BUYING BEVERAGES IN BOTTLES INSTEAD OF CANS... WALKING, TAKING PUBLIC TRANSIT, OR CAR-POOLING INSTEAD OF DRIVING YOUR OWN CAR... TURNING THE THERMOSTAT DOWN IN WINTER, ETC...

在输出方面：

涉及作一些节能的选择，如买瓶装饮料，而不买易拉罐装的；选择步行、乘坐公共交通工具或者拼车，而不开自己的车；冬天将空调温度调低等。

THESE ARE PERSONAL, INDIVIDUAL CHOICES, BUT AS ENERGY PRICES RISE THEY TURN INTO SOCIAL POLICY AND BUSINESS DECISIONS: DESIGNING COMMUNITIES WITH HOUSING CLUSTERS NEAR BUSINESS AND SHOPS, SUBSIDIES FOR MASS TRANSIT, DAYCARE IN THE WORKPLACE TO REDUCE PARENTS' DRIVING, ETC.

这些是个人的选择，但是随着能源价格的上涨，个人选择可以转化为社会政策和商业决策。如在设计社区时，将住宅建在靠近企业和商店的位置，补贴公共交通费用，在工作单位建立日托服务中心以减少父母的开车需要等。

waste:

REDUCING WASTE MEANS INCREASING THE SYSTEM'S EFFICIENCY.

在废料方面：

减少**废料**意味着提高系统的能量效率。

SOURCE REDUCTION MEANS AVOIDING WASTE BY USING ALTERNATIVE PRODUCTS: CLOTH SHOPPING BAGS INSTEAD OF PAPER, TEA INSTEAD OF CANNED DRINKS, ETC.

RECYCLING DIVERTS WASTE ENERGY BACK INTO THE SYSTEM AS INPUT. RECYCLING AN ALUMINUM CAN MEANS THE ENERGY OF MANUFACTURE ISN'T CONSUMED IN ONE USE.

源头减量指通过使用替代产品来避免制造废料：如使用布质购物袋而不使用纸袋，喝茶而不喝罐装饮料等。

回收利用指将废料回收后重新投入输入系统。回收一个铝罐意味着在生产过程中使用的能源不是一次性被消耗掉。

THERE ARE MANY, MANY OTHER WAYS TO INCREASE ENERGY EFFICIENCY, SUCH AS MORE EFFICIENT PRODUCTS, LIKE HIGH-MILEAGE AUTOS OR BETTER-INSULATED HOMES, OR REFRIGERATORS WITH FANS TO COOL THE COILS, OR FURNACES THAT CAPTURE CHIMNEY HEAT, OR BETTER LUBRICANTS TO REDUCE FRICTION.

还有很多提高能源效率的办法，例如生产并使用能效高的产品，如省油的汽车、隔热性好的房子、带冷却线圈风扇的冰箱、可以捕捉烟囱热量的火炉、能够减少摩擦的优质润滑油等。

THERE ARE ALSO INEFFICIENCIES OF THE WHOLE SYSTEM: TOO MANY CONVERSION STEPS, BADLY DESIGNED TRANSPORTATION SYSTEMS, ETC. THESE ARE HARD PROBLEMS TO SOLVE, AS IN SAN FRANCISCO, A CITY WHICH HAS NEVER BEEN ABLE TO BRING MASS TRANSIT ALL THE WAY TO THE AIRPORT!

整个系统也有低效的地方，如太多的转换步骤、设计不合理的运输系统等。这些问题很难解决，如在旧金山，就从来没有设计城市到机场的公共交通线路。

input:

YOU MIGHT SAY THE INPUT WILL TAKE CARE OF ITSELF. WHEN THE OIL RUNS OUT, ENERGY USE IS BOUND TO GO DOWN!

在输入方面：

你可能会说输入会自我调节。石油用光了，能源使用也一定会降低了。

TO MAKE THIS TRANSITION FROM PETROLEUM USE AS SMOOTH AS POSSIBLE, IT'S NECESSARY TO DO WHAT HUMANITY HAS HISTORICALLY DONE: CONSIDER ALTERNATIVE ENERGY SOURCES.

为了使石油使用的转变尽可能顺利，很有必要去做人类祖先曾做过的事情：考虑使用替代能源。

AT ONE TIME,

NUCLEAR POWER

LOOKED LIKE "THE" ANSWER. IN A NUCLEAR REACTOR, RADIOACTIVE FUEL RODS GLOW HOT, DRIVING A HIGH-TECH STEAM ENGINE THAT CAN GENERATE ELECTRICITY, TURN A SHIP'S SCREW, ETC. BECAUSE IT BURNS NOTHING, A NUCLEAR REACTOR EMITS NO SOOT OR OTHER POLLUTING COMBUSTION BY-PRODUCTS.

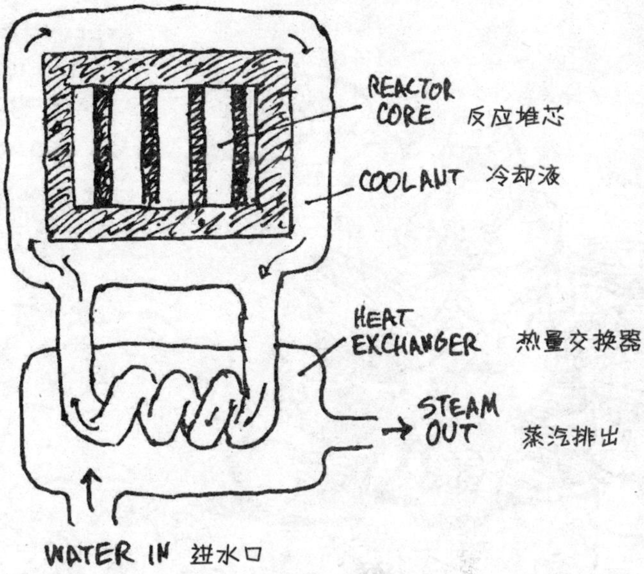

核能曾一度被认为是最好的选择。在核反应堆里，放射性燃料棒发热时，便可以驱动高科技蒸汽机来发电、驱动轮船上的螺旋桨。由于核反应堆不使用任何燃料。因此，也不释放烟灰或者产生其他污染物。

THERE IS ALSO THE PROMISE OF A "BREEDER REACTOR," WHICH CAN ACTUALLY PRODUCE MORE FUEL THAN IT CONSUMES. ENERGY FOREVER!

同时，"增殖反应堆"也有可观的前景。实际上，它产出的燃料比消耗的燃料多很多，从而会有源源不断的能源出现！

BUT OH, THOSE NEGATIVES! RADIATION IS BAD FOR YOUR HEALTH, SO SPECIAL PRECAUTIONS MUST BE TAKEN IN MINING, PROCESSING, AND TRANSPORTING FUEL. IT CAN ALSO BE USED FOR BOMBS, SO YOU HAVE TO GUARD IT! EVERYTHING ABOUT NUCLEAR ENERGY IS VERY EXPENSIVE.

但是，核反应堆也有很多负面影响！辐射对人的健康有害，所以在开采、加工和运输核燃料过程中必须采用特殊的预防措施。同时它还能用于制造核弹，所以必须严加防备！同时，与核能有关的一切都很昂贵。

OOPS!
哎呀!
THEN THERE'S THE QUESTION OF WHAT TO DO WITH THE SPENT FUEL, WHICH STAYS "HOT" BASICALLY FOREVER.
另外还有一个问题，就是如何处理燃烧过的核燃料，废弃燃料可以说永远都是“危险”的东西。
WORST OF ALL, SYSTEM FAILURES CAN REALLY RUIN THE NEIGHBORHOOD...
最严重的是，万一系统出了故障，真的会影响邻近地区……
FOR THE NEXT 10,000 YEARS.
这得影响1万年。
AS I WAS SAYING, "OOPS!"
正如我所说的，“哎呀！”
OOPS AGAIN!
哎呀！还有什么负面影响！
WORLDWIDE, THERE ARE CURRENTLY ABOUT 520 NUCLEAR POWER PLANTS EITHER WORKING OR UNDER CONSTRUCTION, PRODUCING SOME 5% OF THE WORLD'S TOTAL ELECTRICITY... BUT BEYOND THESE, FEW NEW ONES ARE EXPECTED.
目前世界上共有大约520个核电站，包括运行中的和正在建设中的。这些核电站发的电占全世界总电量的5%……但是，估计不会再建造新的核电站了。

SOLAR POWER

FOSSIL FUEL'S ENERGY ULTIMATELY COMES FROM THE SUN. WHY NOT GO DIRECTLY TO THE SOURCE AND ELIMINATE SOME CONVERTORS?

太阳能

化石燃料能源从根本上说来源于太阳。那么，为什么不直接接近其源头，减少一些转换器呢？

PHOTOVOLTAIC CELLS (SOLAR PANELS) CONVERT SUNLIGHT DIRECTLY TO ELECTRICITY. HOWEVER, THEY ARE STILL EXPENSIVE AND INEFFICIENT, MAKING LARGE-SCALE ELECTRICAL GENERATION UNLIKELY FOR SOME TIME. SOLAR HEATING AND HOT WATER ARE FEASIBLE NOW (IN SUNNY PLACES).

光伏电池（太阳电池板）直接将太阳光转化为电。但这种方法很昂贵且效率低，目前不大可能用这种方法大规模发电。现在可以做到的是利用太阳能取暖、供应热水（这也只能在阳光充足的地方）。

HYDROELECTRIC

PLANTS PRODUCE IMMENSE AMOUNTS OF ELECTRICITY, USING THE CLEANEST, CHEAPEST "FUEL" OF ALL: FALLING WATER.

水力发电站产量高、最干净和最便宜的"燃料"非流水莫属了。

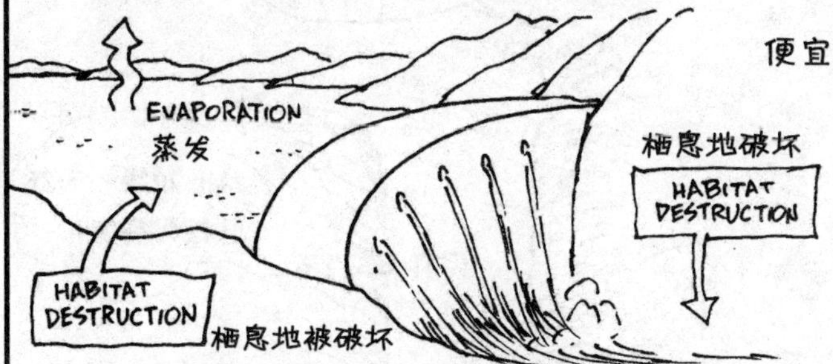

THE DOWNSIDE: THE LARGE DAMS REQUIRED BY HYDROELECTRIC POWER HAVE ENVIRONMENTAL AND SOCIAL COSTS: BEHIND THE DAM, WATER IS LOST THROUGH EVAPORATION, AND FERTILE SILT SETTLES TO THE BOTTOM OF THE LAKE. DOWNSTREAM, REDUCED WATER FLOW DRIES OUT THE ENTIRE WATERSHED.

可其缺点是：水力发电站需要的大坝要付出环境和社会代价，即在大坝后面，水在蒸发过程中损失掉，肥沃的淤泥沉积到湖底。下游水流减少，使整个流域缺水。

WIND

CAN BE USED EITHER TO GENERATE ELECTRICITY OR TO DO WORK DIRECTLY. IN SOME AREAS, WINDMILLS CAN GENERATE ENOUGH POWER TO LIGHT CITIES. BUT WINDMILLS BREAK OFTEN, AND ARE HARD ON BIRDS.

风可以用来发电，也可以直接做功。在一些地区，风车可以产生足够的电力供城市照明。但是风车经常损坏，而且容易伤害鸟类。

BIOMASS-

WOOD, FOR EXAMPLE—IS STILL A VIABLE FUEL. IT'S RENEWABLE, AND CAN BE CONSERVED BY THE USE OF EFFICIENT STOVES. PLANTING FAST-GROWING TREE SPECIES AMELIORATES DEFORESTATION. BUT BURNING BIOMASS IS POLLUTING.

生物质

例如，树木仍然是一种可行的燃料。它可以再生，而且可以使用高效的炉子。种植快速生长的树种可以改善荒漠化。但是燃烧生物质会产生污染。

GEOTHERMAL

ENERGY TAPS THE HEAT OF THE EARTH'S INTERIOR TO MAKE ELECTRICITY, HEAT WATER, OR DO OTHER USEFUL WORK. ALL YOU NEED IS A DEEP HOLE AND SOME CONVERSION TECHNOLOGY. THE HEAT IS DOWN THERE, AND IT'S FREE!

地热能指利用地球内部的热量来发电，供应热水或者发挥其他作用。只需要打一个深孔并利用转换技术就可以了。热量就在地下深处，而且是免费的！

TO SUM UP, THERE REALLY IS AN ENERGY PROBLEM: MODERN LIFE DEPENDS ON LARGE, DAILY INJECTIONS OF ENERGY FROM—***SOMEWHERE.*** AT PRESENT, IT'S FOSSIL FUELS, BUT THEY CAN'T LAST FOREVER.

总之，能源的确是个问题：现代生活每天都依赖从某处注入大量的、**源源不断**的能源。目前是化石燃料，但是化石燃料不能维持到永远。

A TECHNOLOGICAL SOLUTION TO THIS PROBLEM, IF IT EXISTS, WILL PROBABLY HAVE TO INCLUDE MANY INGREDIENTS: LOWER-CONSUMPTION LIFESTYLES IN THE DEVELOPED WORLD, MORE EFFICIENT CONVERTORS, AND SAFE, NON-POLLUTING ALTERNATIVE SOURCES OF ENERGY. THERE ARE INNUMERABLE SPECIFIC AREAS TO WORK ON, FROM TRANSPORTATION TO MORE EFFICIENT COOKSTOVES FOR THE THIRD WORLD...

如果采用技术手段解决这个问题的话（不知道究竟有没有这种技术手段），很可能必须包括许多内容，如发达国家必须采用较低消费的生活方式、更加有效的转换器、安全无污染的替代能源。值得研究的具体领域不计其数，包括从交通运输领域到为第三世界生产的能效更高的烹饪炉。

WILL THE EFFORT PAY OFF?

这些努力会有成效吗？

CHAPTER 11

LET'S EAT AGAIN!

第十一章

让我们继续开饭！

IN THE LAST CHAPTER, WE DESCRIBED ENERGY NETWORKS AS IF THEY WERE SEPARATE FROM FOOD WEBS. IN FACT, THE TWO KINDS OF ENERGY ARE CLOSELY RELATED. IN THE FIRST PLACE, FOOD IS AT THE BASE OF THE WHOLE ENERGY CHAIN. IF NOBODY EATS, NOBODY PUMPS GAS!

在上一章中，我们描述了各种能源网，它们似乎与食物网是互相独立的，而事实上，这两种不同类型的能量是密切相关的。首先，食物处在整个能量链中最基础的位置。如果没有人吃东西，也就没有人需要加油了。

SECOND, MOST FOOD IS GROWN USING CHEMICAL FERTILIZERS, WHICH ARE MADE FROM PETROLEUM PRODUCTS. THAT IS, FOOD ENERGY IS DERIVED FROM FOSSIL FUELS!

其次，大多数食物的生长是需要使用化肥的，而化肥是由石油产品制造出来的。那就意味着食物能量来源于化石燃料！

FINALLY, THE GLOBAL INDUSTRIAL ECONOMY PLAYS A MAJOR ROLE IN DECIDING WHICH CROPS ARE PLANTED AND WHERE THEY ARE SHIPPED.

最后，全球工业经济在很大程度上决定了种植什么农作物和食品以及将其运到何处。

AGRICULTURAL PRACTICES VARY WIDELY FROM COUNTRY TO COUNTRY AND REGION TO REGION. IN THE INDUSTRIALIZED WEST, WE USE FOSSIL-FUEL DEVOURING HEAVY FARM MACHINERY AND PLENTY OF FERTILIZERS AND PESTICIDES, WHILE IN ENERGY-POOR REGIONS, FARMERS RELY ON MUSCLE POWER AND COOK WITH WOOD OR CHARCOAL.

农业生产在国家与国家之间、地区与地区之间是大不一样的。在工业化的西方，人们在农场使用以化石燃料为能源的大型机械、化肥和农药；而在能源缺乏的地区，农民只能依靠体力劳动，并且只能用木柴或木炭来做饭。

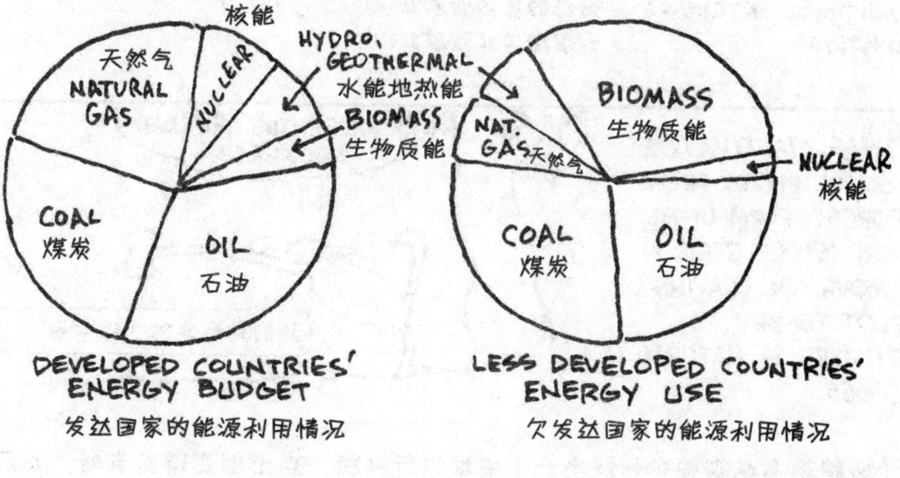

AT PRESENT, ABOUT 25% OF THE ARABLE LAND IS USED FOR ***NOMADIC HERDING***... ANOTHER 25% IS UNDER OLD-FASHIONED ***SHIFTING CULTIVATION***, AND THE REST IS DIVIDED BETWEEN ***TRADITIONAL INTENSIVE AGRICULTURE*** (MOSTLY IN ASIA) AND ***LARGE-SCALE INDUSTRIAL AGRICULTURE*** (EUROPE AND NORTH AMERICA).

现在，世界上25%的农业用地为**游牧业**所用……另外25%为落后的**迁移农业**所用，剩下的50%农业用地为传统的**集约型农业**（主要在亚洲）和**大规模工业型农业**（欧洲和北美）所用。

SHIFTING CULTIVATOR HACK SMALL PLOTS FROM THE FOREST, FARM UNTIL THE SOIL IS DEPLETED, AI THEN MOVE ON, LEAVING THE PLOT TO BE REFERTILIZED BY NATURAL PROCESSES.

迁移种植者从森林中开辟出一小块地进行种植，当土壤变得贫瘠时，他们就迁移到别处，让原来的土地通过自然的过程重新变得肥沃。

TRADITIONAL INTENSIVE FARMING PRODUCES VERY HIGH CROP YIELDS BY USING LOTS OF LABOR AND RECYCLING NUTRIENTS EFFICIENTLY. CHINESE FARMERS, WHO CYCLE NUTRIENTS THROUGH PIGS, CHICKENS, DUCKS, FISH, AND ALGAE, RETURNING ORGANIC MATTER TO THE SOIL, ACHIEVED OUTPUT ROUGHLY ***TEN TIMES*** THAT OF PREINDUSTRIAL EUROPE. CHINA HAS MAINTAINED REASONABLY GOOD SOIL FERTILITY FOR MILLENNIA.

传统的集约型农业通过投入大量的劳动力和有效地循环利用养分使得农作物的产量非常高。中国农民通过猪、鸡、鸭、鱼和藻类来循环利用养分，并将产生的有机物重新投入土地里，使农作物产量达到欧洲工业化前农作物产量的10**倍**左右。这种耕作方式使中国的土地几千年来一直相当肥沃。

INDUSTRIAL AGRICULTURE, WHICH RELIES ON HEAVY MACHINERY AND CHEMICALS, IS ***HIGH-INPUT AGRICULTURE.*** IT DRAWS ENERGY FROM FOSSIL FUELS AND CONVERTS IT TO FOOD OR OTHER CROPS. (BUT DON'T FORGET THAT AT HARVEST TIME, FRUITS AND VEGETABLES ARE MOSTLY PICKED BY VERY LOW-PAID TEMPORARY WORKERS.)

工业型农业是一种**高投入农业**，它依赖于大型机械和各种化学品的投入。这种农业从化石燃料中获得能量，并把这种能量转化成食物或者其他农产品（但不要忘了，在收获季节，所有水果和蔬菜都是由工资很低的临时工来采摘的）。

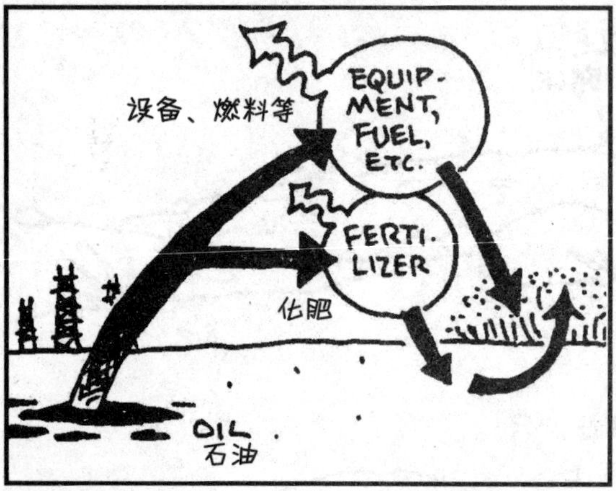

AN INDUSTRIAL FARM NEEDS LARGE AMOUNTS OF CAPITAL (A FANCY NAME FOR MONEY), AMPLE ACREAGE, AND CHEAP ENERGY. LIKE ALL INDUSTRY, INDUSTRIAL AGRICULTURE GREW UP IN THE ERA OF PLENTIFUL FOSSIL FUEL.

工业型农业需要大量的资本（金钱的一种美称）、充足的土地和便宜的能源。像所有的工业一样，工业型农业也是在能源丰富的时代成长起来的。

WHEN FUEL IS CHEAP, THERE IS LITTLE INCENTIVE TO CONSERVE IT, SO INDUSTRIAL FARMING HAS BEEN ALMOST UNBELIEVABLY WASTEFUL. GRAIN-FED CATTLE, FOR EXAMPLE, CONSUME AROUND *7 TIMES* AS MUCH ENERGY AS THEY PRODUCE AS BEEF.

当燃料便宜的时候就没有保存的必要了，因此工业型农业向来是非常浪费的。例如，用谷物饲养的牛消耗的能量是其牛肉产生能量的7倍。

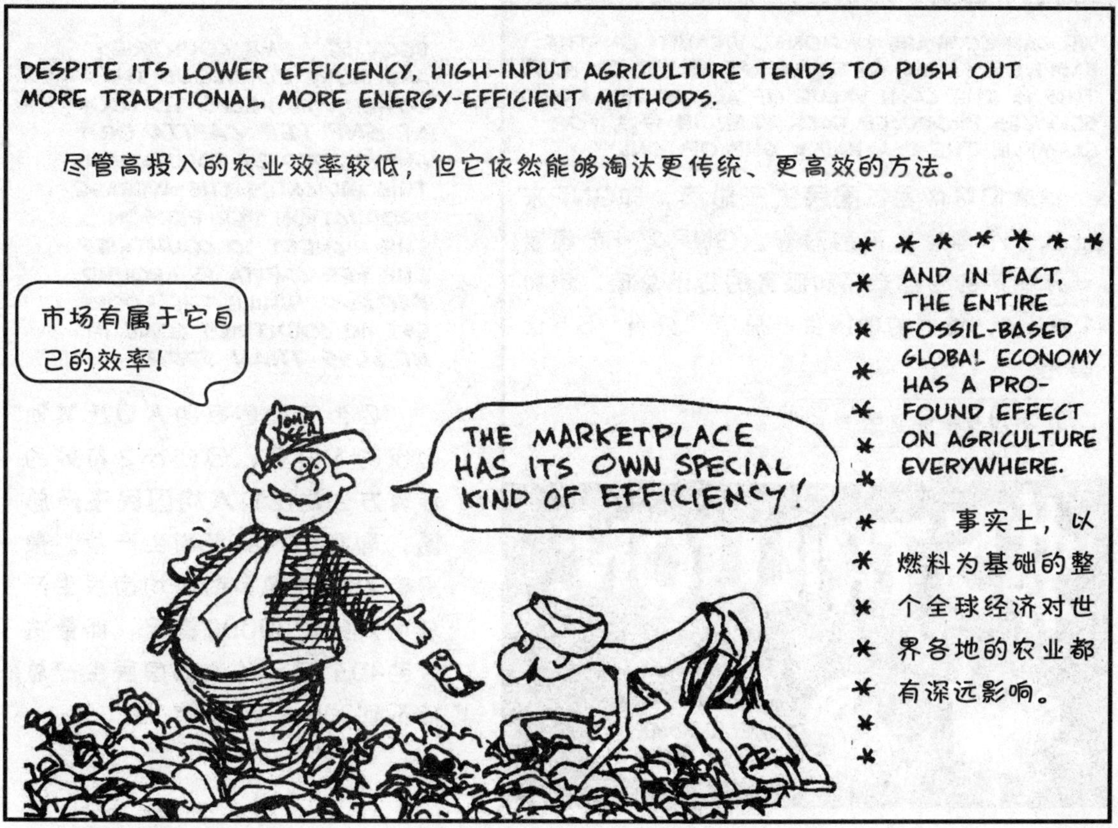
DESPITE ITS LOWER EFFICIENCY, HIGH-INPUT AGRICULTURE TENDS TO PUSH OUT MORE TRADITIONAL, MORE ENERGY-EFFICIENT METHODS.
尽管高投入的农业效率较低，但它依然能够淘汰更传统、更高效的方法。
市场有属于它自己的效率！
THE MARKETPLACE HAS ITS OWN SPECIAL KIND OF EFFICIENCY!
AND IN FACT, THE ENTIRE FOSSIL-BASED GLOBAL ECONOMY HAS A PROFOUND EFFECT ON AGRICULTURE EVERYWHERE.
事实上，以燃料为基础的整个全球经济对世界各地的农业都有深远影响。

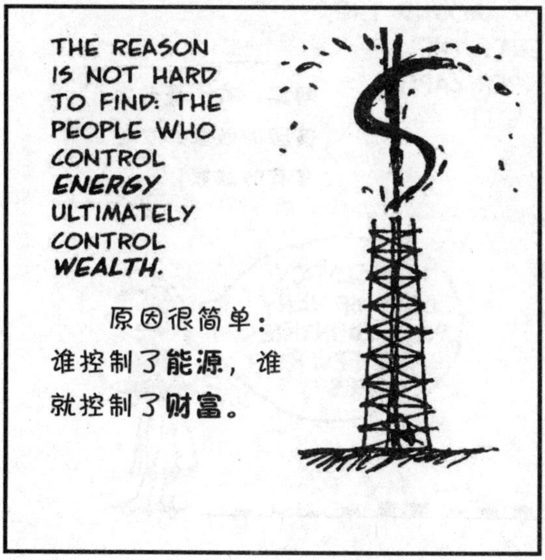
THE REASON IS NOT HARD TO FIND: THE PEOPLE WHO CONTROL ENERGY ULTIMATELY CONTROL WEALTH.
原因很简单：谁控制了能源，谁就控制了财富。

THE RICHEST COUNTRIES—AND THE RICHEST COMPANIES—CAN CONTROL MARKETS, TECHNOLOGY, MINERAL RESOURCES, AND LAND USE.
$
最富裕的国家和公司能够控制市场、技术、矿藏和土地的使用。

WE CAN COMPARE NATIONAL WEALTH ON THE BASIS OF *GROSS NATIONAL PRODUCT*, OR *GNP*. THIS IS THE CASH VALUE OF ALL GOODS AND SERVICES PRODUCED EACH YEAR. IN 1993, FOR EXAMPLE, THE USA HAD A GNP OF "ONLY"

我们可以通过**国民生产总值**，即GNP来比较各个国家之间的财富。GNP是一个国家一年生产的全部产品和服务的货币总值。例如1993年，美国的国民生产总值“只有”5万亿美元。

BECAUSE SOME COUNTRIES ARE MORE POPULOUS THAN OTHERS, IT HELPS TO LOOK AT ***GNP PER CAPITA*** OR GNP DIVIDED BY POPULATION. THIS INDICATES THE AVERAGE PRODUCTION PER PERSON. THE RICHEST 10 COUNTRIES' GNP PER CAPITA IS AROUND ***$20,000,*** WHILE THE POOREST 40 COUNTRIES COME IN AT ***LESS THAN $300.***

因为有些国家的人口比其他国家的人口多，因此行之有效的计算方法是比较**人均国民生产总值**，即每个人的平均生产量。最富有的10个国家的人均国民生产总值大约是20000**美元**，而最贫穷的40个国家的人均国民生产总值**不到**300**美元**。

THIS GRAPH SHOWS HOW WEALTH IS SPREAD AROUND THE WORLD. THE HEIGHT OF EACH BAR REPRESENTS THE NUMBER OF COUNTRIES WITH A GIVEN GNP PER CAPITA.

这张图显示的是财富在世界上的分布情况。每个竖条的高度代表了给定的人均国民生产总值的国家数。

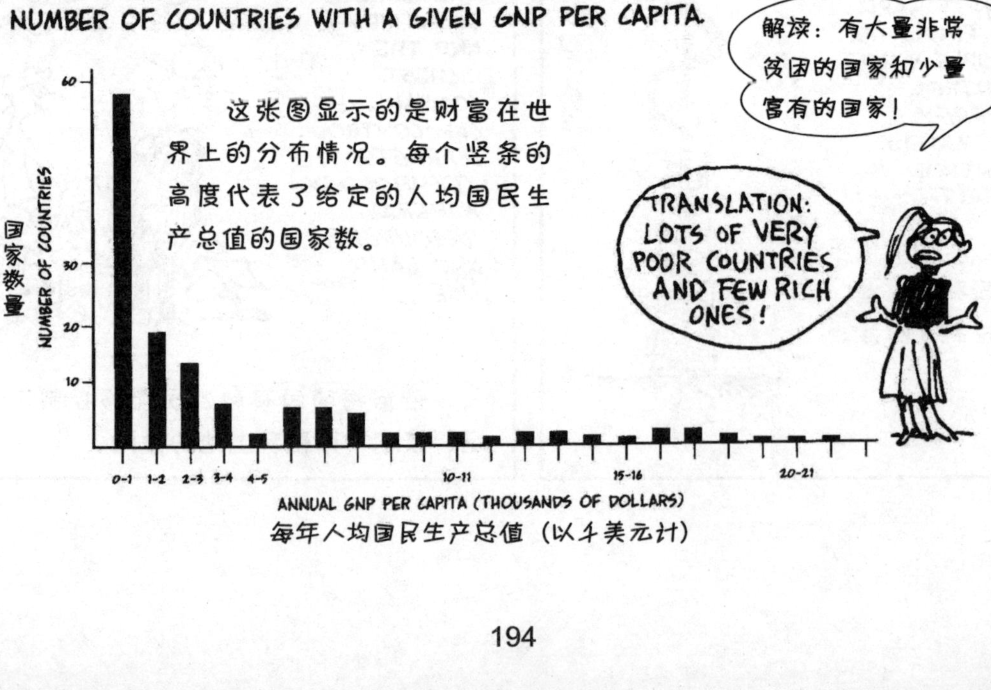

BECAUSE INDUSTRIALIZATION HAS BEEN THE KEY TO WEALTH, NEARLY EVERY COUNTRY ON EARTH WANTS ***MORE INDUSTRY.*** BUT HOW TO GET IT? THE BUSINESSES OF THE WEST AREN'T GIVING IT AWAY!

因为工业化是获得财富的关键，因此世界上几乎每一个国家都想要**更多的工业**。可怎样才能有更多的工业呢？西方国家的企业是不会把工业拱手相送的：

THE USUAL WAY IS FOR POOR COUNTRIES TO SELL ***AGRICULTURAL PRODUCTS*** ON THE WORLD MARKET, AND USE THE MONEY TO BUY OIL AND MACHINERY.

对于贫穷的国家来讲，常用的方式是在世界市场上将**农产品**卖掉，然后用得到的钱购买石油和机器。

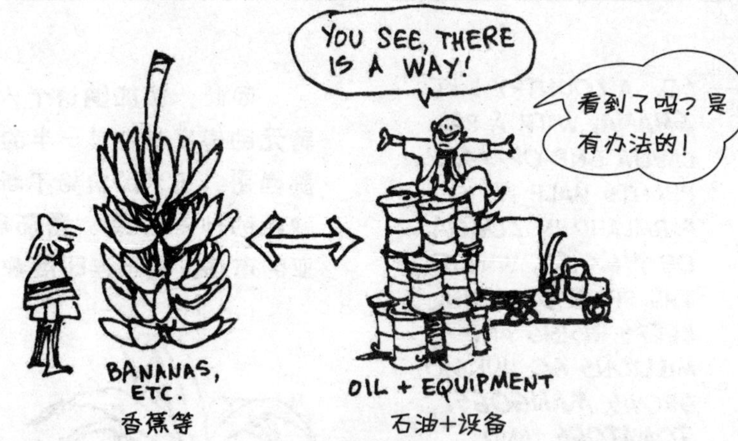

BUT THE WORLD MARKET ISN'T ESPECIALLY INTERESTED IN LOCAL FOOD CROPS LIKE MANIOC, TAPIOCA, YAMS, OR BREADFRUIT. THE WORLD WANTS CHOCOLATE, COFFEE, BANANAS, CASHEWS, AND PISTACHIO NUTS, OR NON-FOOD PLANT PRODUCTS LIKE COPRA, JUTE, AND RUBBER...

可是世界市场对木薯、木薯粉、山药或面包果等土特产并不是特别感兴趣。世界市场感兴趣的是可可、咖啡、香蕉、腰果、阿月浑子果，还有椰肉、黄麻、橡胶等非食品类植物产品……

RESULT: LANDOWNERS PLANT THESE ***"CASH CROPS"*** ON LAND THAT MIGHT HAVE BEEN GROWING FOOD INSTEAD.

结果是：土地所有者在原本应该种植粮食作物的土地上种植这些**"经济作物"**。

SO... A COUNTRY LIKE ***GHANA,*** WITH A PER CAPITA GNP OF ***$400,*** PLANTS HALF ITS FARMLAND IN ***COCOA...*** OR ***MEXICO,*** WHERE THE PRICE OF CORN KEEPS RISING AND MILLIONS GO HUNGRY, GROWS ***MANGOES, TOMATOES,*** AND ***MELONS*** FOR CALIFORNIA. EVEN ***INDIA*** EXPORTS ***RICE!***

因此，像**加纳**这个人均国民生产总值只有400**美元**的国家却在其一半的农田上种植了**可可**……在**墨西哥**，玉米的价格不断上涨，数百万人挨饿，而种植的却是**芒果**、**番茄**和**甜瓜**，来供应加利福尼亚的市场。甚至连**印度**都出口**大米**！

IN THE 1960s AND 1970s, AGRICULTURAL SCIENTISTS TRIED TO ADDRESS THE PROBLEM OF WORLD HUNGER BY CREATING THE **GREEN REVOLUTION.**

(IN THOSE DAYS, EVERYTHING WAS A REVOLUTION!)

20世纪60～70年代，农业科学家尝试通过**“绿色革命”**来解决世界的饥饿问题。（那个年代什么都是革命！）

THEY BRED VERY HIGH-YIELD VARIETIES OF WHEAT, RICE, AND MILLET, WITH HEAVY HEADS OF GRAIN, SHORT, TOUGH STEMS TO SUPPORT THEM, AND A SHORT GROWING SEASON (ALLOWING DOUBLE CROPPING). THESE TYPES COULD RAISE CROP YIELDS BY ***TWO TO THREE TIMES.***

科学家种植的小麦、大米和谷子都是高产量的品种：谷穗饱满、秆茎矮壮、生长期很短（一年可以种植两季）。这些品种的作物可以将产量提高**两到三倍**。

MANY POOR COUNTRIES EMBRACED THE GREEN REVOLUTION... AND THEN MARKET FORCES TOOK OVER. IT TURNED OUT THERE WAS A DOWNSIDE TO THE NEW, REVOLUTIONARY GREENS: THEY NEEDED A LOT OF FERTILIZER AND WATER. THEIR HIGH OUTPUT DEPENDED ON HIGH INPUT!

许多贫穷国都非常欢迎这场绿色革命……但后来市场的力量占了上风。殊不知，这种刚兴起的绿色革命也有其缺点：这种新品种需要大量的化肥和水，高产出依赖于高投入！

ONLY A FEW AFFLUENT FARMERS COULD AFFORD THE FERTILIZER, ETC. THEY PLANTED THE NEW VARIETIES... THEIR LAND PRODUCED MORE... GRAIN PRICES DROPPED... AND THEIR POOR NEIGHBORS, STUCK WITH THE OLD VARIETIES, BECAME EVEN POORER.

只有少数有钱的农民才能支付得起这么多的化肥等投入。他们种植最新品种的作物，粮食产量增加了……价格下降了……而他们贫穷的邻居仍然种植老品种，因而变得更加贫困。

DISCLAIMER: THE OPINIONS EXPRESSED BY CHARACTERS IN THIS BOOK ARE NOT NECESSARILY THOSE OF THE AUTHORS.

声明：本书人物中表达的观点与本书作者及出版社无关。

THE SUCCESSFUL ONES BOUGHT MORE LAND, ENLARGING THEIR HOLDINGS AT THE EXPENSE OF THEIR NEIGHBORS. LANDLESS FARMERS FLOCKED TO THE CITIES, HOPING FOR JOBS—IN AN INDUSTRIAL ECONOMY THAT BARELY EXISTED YET.

成功的农民购买了更多的土地，他们在牺牲邻居利益的基础上，扩大了自己的土地占有量。于是，失去土地的农民大量涌入城市寻找工作——这种情况在工业化社会里已经很少存在了。

RESULT: MORE INDUSTRIAL AGRICULTURE, MORE FOOD, AND MORE POVERTY!

结果是：出现了更多的工业型农业，生产出了更多的粮食，也出现了更多的穷人！

IN INDIA, 75% OF FARMERS ARE POORER NOW THAN THEY WERE IN 1975, EVEN THOUGH INDIA PRODUCES MORE FOOD PER PERSON NOW THAN IT DID THEN.

虽然现在印度的人均粮食产量要比1975年高，但是仍有75%的农民要比那时候贫困。

THE CHANGES SINCE 1950 HAVE BEEN OVERWHELMING:

自1950年以来变化是巨大的：

- TOTAL CROPLAND INCREASED **22%** (FROM 2.3 TO 2.8 MILLION SQ. MI.)
- CHEMICAL FERTILIZER CONSUMPTION UP **TENFOLD** (OVER 20 KILOS PER PERSON ON EARTH ANNUALLY)
- IRRIGATION AREA TRIPLED
- AGRICULTURAL CONSUMPTION OF OIL UP TO 1/12 THE WORLD'S TOTAL PRODUCTION.

- 农田面积增加了22%（从230万平方英里增长到了280万平方英里）
- 化肥消费量增长了10倍（每年全世界人均消费二十多千克）
- 灌溉面积增加到原来的3倍
- 农业生产的石油消费量增长到全世界石油总产量的1/12。

WITH THE OVERALL RESULT THAT TOTAL WORLD FOOD PRODUCTION HAS **TRIPLED** IN THAT TIME.

最终结果是现在全世界的粮食产量是1950年的3倍。

SINCE POPULATION HAS "ONLY" DOUBLED IN THE SAME PERIOD, THERE IS MORE FOOD PER PERSON NOW THAN AT ANY TIME IN THE PAST 50 YEARS...

由于同一时期人口“只”翻了一番，因此，现在的人均粮食产量要比过去的50年里的任何一段时间都多。

BUT WE'VE SEEN THE COSTS: ALTHOUGH THERE'S MORE FOOD FOR EVERYONE TO EAT, "EVERYONE" ISN'T EATING IT. INSTEAD, WE HAVE GROWING MIDDLE AND UPPER CLASSES, WHOSE STANDARD OF LIVING HAS RISEN SHARPLY, WHILE MOST OF THE POPULATION HAS LESS.

但是我们也看到了代价：尽管有了更多的粮食，但并不是“所有人”都有的吃。相反，中产阶级和上层阶级的财富增长，生活水平迅速提高，而大部分人并不富裕。

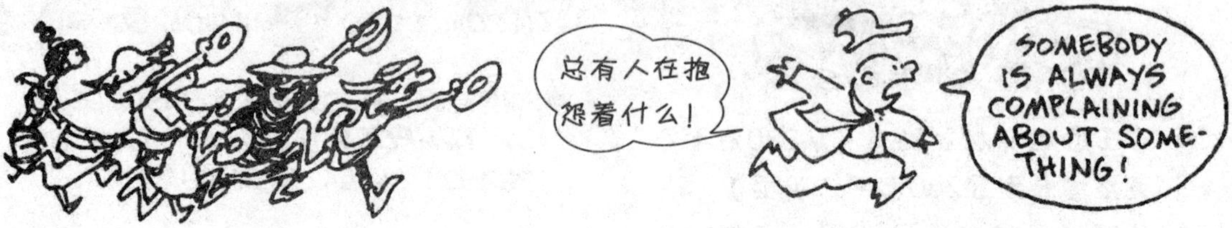

MEANWHILE, TRADITIONAL AGRICULTURE IS LOSING GROUND, AND THE SYSTEM DEPENDS MORE AND MORE ON HIGH INPUTS. CAN THIS BE SUSTAINED?

同时，传统农业渐渐消失，农业生产越来越依赖于高投入。这种情况能维持下去吗？

ASIDE FROM THE LIMITS ON WORLD PETROLEUM RESERVES, THE ***PHOSPHORUS*** CYCLE ALSO LIMITS INDUSTRIAL AGRICULTURE. CHEMICAL FERTILIZERS USE PHOSPHATES DUG FROM MINES, WHICH SHOULD PLAY OUT AROUND THE YEAR 2050 AT PRESENT RATES OF CONSUMPTION.

除了世界石油储藏量的制约外，磷的循环也同样制约着工业型农业的发展。生产化肥需要从矿石中得到磷，而以目前的消费速度来看，这些磷将在2050年左右用完。

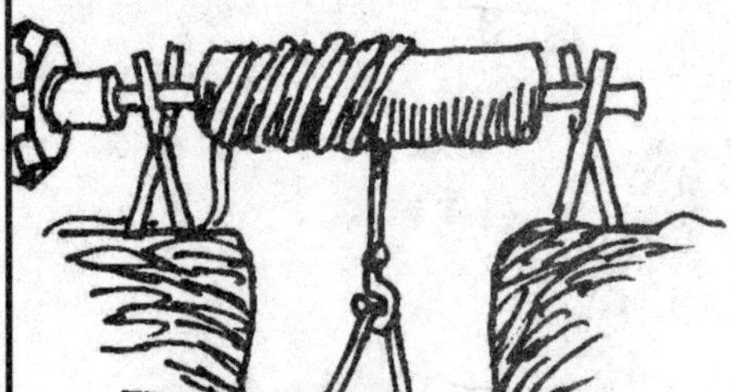

(TRADITIONAL INTENSIVE AGRICULTURE GETS ITS PHOSPHATES FROM FISH, BONE MEAL, AND RECYCLED WASTE.)

（传统的精耕细作型农业从鱼、骨粉和能够循环利用的废物中获得磷。）

THESE CHANGES TAKE THEIR TOLL ON THE NATURAL ECOSYSTEMS AND THEIR DIVERSITY.

这些变化给生态系统及其多样性带来了负面影响。

EVERYTHING WE SAID IN CHAPTER 7 ABOUT FARMING'S EFFECTS ON WILD SYSTEMS GOES DOUBLE FOR INDUSTRIAL AGRICULTURE.

在第七章我们探讨过农业生产对生态系统的影响，工业型农业对生态系统的影响是传统农业对生态系统影响的两倍。

INDUSTRIAL GROWERS USUALLY RAISE JUST ONE OR TWO VARIETIES OF WHEAT OR CORN, APPLES OR PEACHES, INSTEAD OF THE HUNDREDS OF LOCAL VARIETIES THAT USED TO THRIVE.

工业型农业的种植者常常只种植小麦、玉米、苹果或梨的一两个品种，而不是像以前那样在当地有上百个不同的农产品品种蓬勃生长。

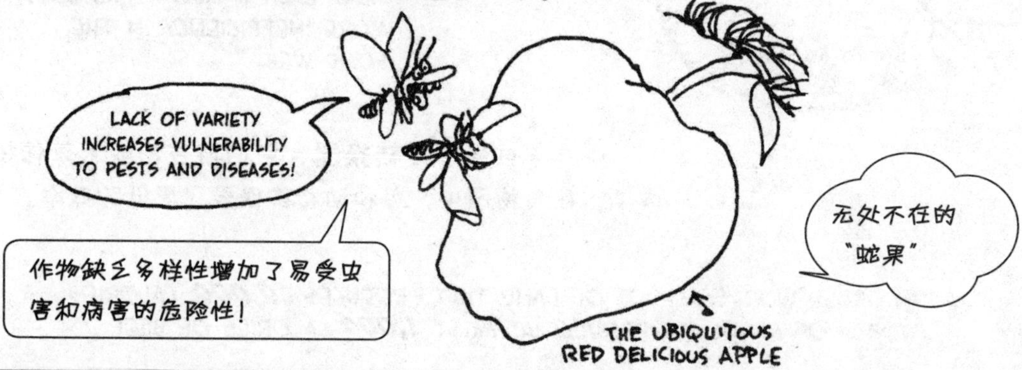

MEANWHILE, MOST OF THE WORLD'S POTENTIAL FARMLAND LIES IN THE TROPICAL RAIN-FOREST, WITH ITS COMPLEX ECOLOGY AND POOR SOIL. THE RESULT IS AN IRRETRIEVABLE LOSS OF BIODIVERSITY, AND AN EROSION OR EXHAUSTION OF TOPSOIL.

同时，世界上大部分可开垦的耕地都位于热带雨林地区。这些地区生态多样，土壤贫瘠。在这些地区开垦土地，会造成生物的多样性消失、不可恢复以及表层土的严重损耗。

AND THEN THERE'S THE ISSUE OF MEAT.

然后是关于肉的问题

A COW OR A SHEEP OR A PIG IS A ***CONVERTOR***, CHANGING GRAIN OR GRASS ENERGY INTO MEAT. EACH ANIMAL REPRESENTS MORE INEFFICIENCY IN THE FOOD WEB.

牛、羊或猪都是**转换器**，它们将谷物或禾草转化成肉类。在食物网中，每种动物都代表了更低的效率。

IN OTHER WORDS, A PLOT OF LAND THAT PRODUCES ***10,000*** CALORIES WORTH OF GRAIN CAN PRODUCE AT MOST ***1,000*** CALORIES OF BEEF.

换句话说，一块土地用来种植谷物，其谷物能产生10000卡*的能量。而用这些谷物喂养牛，其牛肉只能产生1000卡的能量。

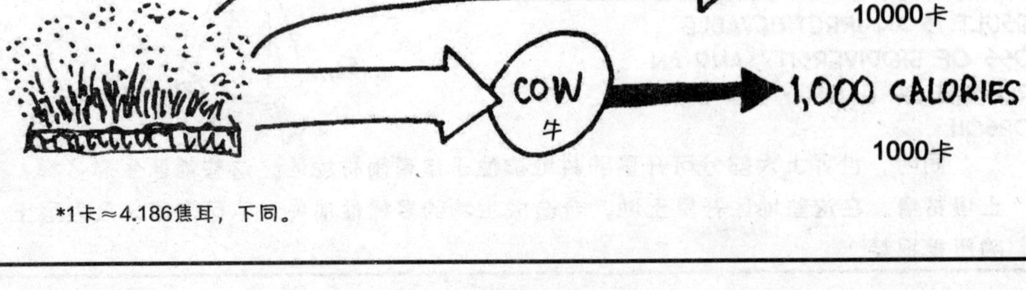

*1卡≈4.186焦耳，下同。

AS PEOPLE GROW MORE AFFLUENT, THEY EAT MORE MEAT. AT PRESENT, TOTAL WORLD PRODUCTION OF GRAIN STANDS AT 2150 MILLION TONS, BUT ABOUT **40%** GOES INTO (NONHUMAN) ANIMAL MOUTHS. ANOTHER MAJOR CONVERTOR OF GRAIN IS **BEER**...

当人们越来越富裕时，他们食用的肉也越来越多。目前，世界上的粮食总产量是21.5亿吨，可有40%的粮食用来喂养动物了（不是被人食用）。另一种主要的粮食转化者是**啤酒**……

AS WE'VE SEEN, HUMANS TAKE ALMOST 40% OF THE LAND'S NET PRIMARY PRODUCTION, BUT ONLY 3% IS ACTUAL CROP. THE REST IS FOREST CLEARING, SETTLEMENT BUILDI NG, CHAFF, AND WASTE.

正如我们在前面所看到的，人类只占有土地净初级生产量的39%，其中，真正的作物只占3%。其余的则是森林砍伐、建筑、谷糠和废物。

AS POPULATION RISES TOWARD 6, 7, 8 BILLION, HUMANITY WILL SEEK WAYS TO WRING MORE FOOD FROM THE EARTH, AND ONE WAY IS SURELY TO IMPROVE EFFICIENCY. IF WE COULD RAISE THAT 3% TO 6%, THERE WOULD BE PLENTY OF FOOD.

当人口增长到60亿、70亿甚至80亿的时候，人类会想办法从土地上获得更多的食物，毫无疑问，提高生产效率是一种有效方式。如果能够从3%提高到6%，那么将会生产出大量的粮食来。

HOW IS THAT TO BE DONE? BY DOUBLE-CROPPING, CAREFUL LAND USE, MORE LABOR-INTENSIVE CONSERVATION PRACTICES ON INDUSTRIAL FARMS, LESS SPOILAGE, AND ***FEWER HAMBURGERS.***

怎么才能做到提高粮食生产呢？方法是种植双季作物，谨慎利用土地，在工业化的农场上利用劳动力精耕细作，减少对生态系统的破坏，还要**少吃汉堡**。

BIOTECHNOLOGY ALSO HOLDS OUT SOME HOPE, FOR EXAMPLE, IN PRODUCING DISEASE-RESISTANT PLANT VARIETIES, AND EVEN IMPROVING PLANT AND ANIMAL CONVERSION EFFICIENCY.

生物技术也同样给我们带来了希望。例如，研制出抗病虫害的植物品种，提高植物和动物的转化效率。

ANOTHER APPROACH IS LOCAL

EMPOWERMENT.

IT'S HARD—IMPOSSIBLE—TO SOLVE THESE PROBLEMS FROM AFAR. ACTION IS MORE EFFECTIVE WHEN TAKEN BY THE PEOPLE DIRECTLY ON THE SPOT.

另一种途径是**赋予当地人们决策权**。

通过他人来解决这些问题是困难甚至是不可行的。当地的人们直接采取措施会更有效。

BUT THEY NEED HELP.

可是他们需要帮助。

A CURRENT EFFORT INVOLVES SPREADING THE USE OF MORE EFFICIENT ***COOKING STOVES***, WHICH PRODUCE THE SAME HEAT FROM MUCH LESS WOOD.

当前的一种方法是努力推广使用效率更高的**炉灶**，这种炉灶用更少的木柴便可以产生同样的热量。

THE RATIONALE: IF POOR RURAL AFRICAN WOMEN DIDN'T HAVE TO SPEND 6-8 HOURS A DAY COLLECTING WOOD, THEY MIGHT HAVE MORE TIME TO DO OTHER WORK—AND DEFORESTATION WOULD SLOW, AS WELL.

合理的解释是：如果非洲农村的一些贫困妇女每天不用花6~8个小时来捡拾木柴，那么她们就会有更多的时间来做其他的工作。同时，砍伐森林的速度也会放缓。

WHAT DOES THIS HAVE TO DO WITH AGRICULTURE? ***FORESTS CAN BE AN ESSENTIAL PART OF THE AGRICULTURAL ENERGY WEB.*** THEY CATCH MOISTURE AND GENERATE NUTRIENTS THAT TRICKLE DOWNSTREAM.

这与农业又有什么必然联系呢？**森林是农业能量网中重要的一部分。**森林获得水分并向地下输送养分。

AND FINALLY, WE NEED TO SAY A WORD ABOUT FISH. SEAFOOD HAS ALWAYS BEEN A TASTY, CONVENIENT SOURCE OF ANIMAL PROTEIN, AND IT MIGHT SOUND LIKE A GOOD ALTERNATIVE TO BEEF, PORK, OR LAMB.

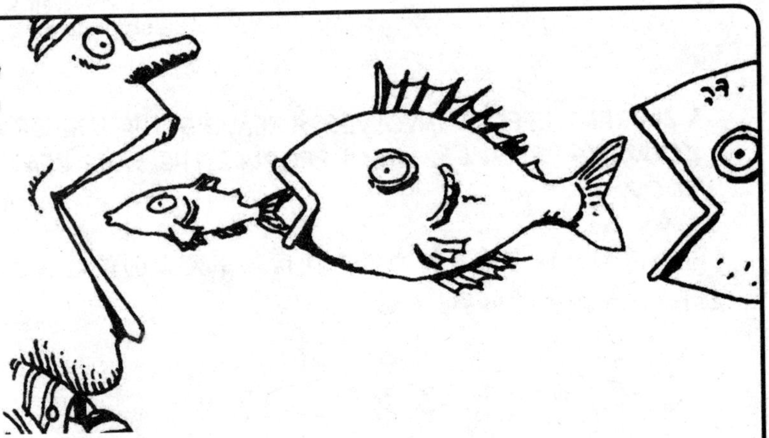

最后，我们需要说一下鱼的问题。海产品一直是一种可口的、方便的动物蛋白质来源，听起来鱼或许是牛肉、猪肉和羊肉的一种很好的替代品。

UNFORTUNATELY, FISHING HAS ALSO TURNED INDUSTRIAL, AND DEMAND IS HUGE. TO FEED A HUMAN POPULATION IN THE BILLIONS, FISHING FLEETS HAVE NOW EXHAUSTED MANY OF THE WORLD'S PRIME FISHING GROUNDS. FISH POPULATIONS, ESPECIALLY THE PREFERRED KINDS, ARE FALLING, SO HUNTERS ARE TURNING TO SMALLER FRY AND SPECIES THAT WERE PREVIOUSLY SCORNED.

可惜渔业也形成了工业化，而且需求量巨大。为了满足几十亿人口的需要，大批渔船出海捕鱼，使得世界上许多主要渔产区渔业资源匮乏。鱼的数量尤其是那些人们喜欢吃的品种的数量开始下降，因此渔民开始转向捕捞小鱼以及捕捞那些以前人们不屑一顾的鱼类。

USES 15 CALORIES OF FUEL TO ACQUIRE 1 CALORIE OF FISH.
用15卡热量的燃料捕获1卡热量的鱼。

THE COMMITTEES THAT "REGULATE" FISHING HAVE USUALLY BEEN DOMINATED BY THE FISHING INDUSTRY... AND GOVERNMENTS ARE INCREASINGLY DOMINATED BY NON-FARMING CITY DWELLERS, AS CITIES SWELL WITH INDUSTRIALIZATION. IN OUR NEXT CHAPTER, WE LOOK AT URBAN ECOLOGY...
“监管”捕鱼的委员会常常被渔业公司控制……随着城市因工业化而膨胀，政府部门开始逐渐受不从事农业的市民支配。在下一章中，我们将会探讨城市生态学。
MEANWHILE, LET'S FINISH WHAT'S ON OUR PLATES...
同时，我得把盘子里的食物吃完……

CHAPTER 12

BRIGHT LIGHTS, BIG CITY

◆第十二章◆

光影大都市

ENERGY CONSUMPTION HAS FUELED INDUSTRY... ENERGY CONSUMPTION HAS SWELLED POPULATION... ENERGY CONSUMPTION HAS CHANGED AGRICULTURE... AND ENERGY CONSUMPTION HAS INCREASED ORGANIZATION...

能源消耗加速了工业化进程……

能源消耗膨胀了人口数量……

能源消耗改变了农业发展方式……

能源消耗增进了组织化程度……

IN OTHER WORDS, PEOPLE HAVE BEEN MOVING OFF THE FARM AND INTO THE CITIES...

换句话说，人们正逐步脱离农场，转而向城市进军……

CITIES PLAY MANY ROLES IN THE ECOSPHERE... WE CAN SUMMARIZE THEM ALL BY SAYING THAT A CITY IS WHERE ***SURPLUSES OF ENERGY AND MATERIAL ARE CONCENTRATED AND TRANSFORMED.***

城市在生态圈中扮演着众多重要角色……这些角色可以总结为一句话，即城市就是**过剩的能源与材料集中和转化**的地方。

ON THE ONE HAND, CITIES ARE MARKETPLACES, MANUFACTURING CENTERS, AND DISTRIBUTION HUBS. GOODS FLOW IN... FACTORIES HUM... PEOPLE BUY AND SELL... AND GOODS FLOW OUT...

一方面，城市是市场、制造中心和分销中心。货物运入城市……工厂忙着加工……人们忙着买卖……货物再运出城市……

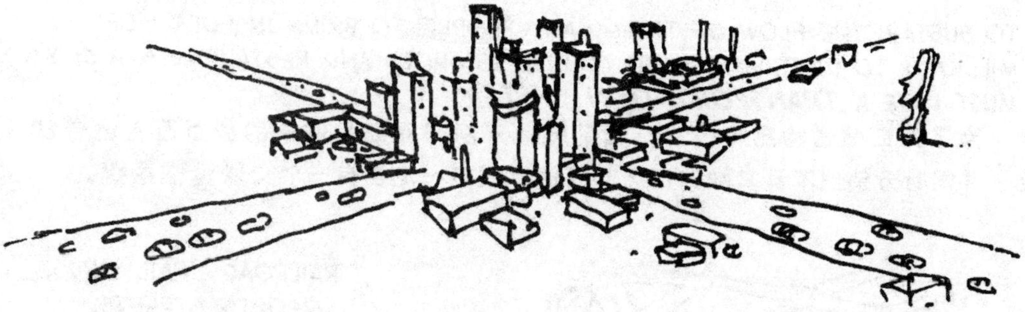

CITIES ARE ALSO CENTERS OF ***ORGANIZATION.*** ENERGY FLOWS IN TO FEED, HEAT, HOUSE, ENTERTAIN, AND MAKE CONNECTIONS BETWEEN PEOPLE WHO MOVE INFORMATION AROUND. THESE INCLUDE ADMINISTRATORS AND MANAGERS, BOTH CORPORATE AND CIVIC, PLUS EVERYONE INVOLVED IN GUIDING THOUGHT AND OPINION, FROM RELIGIOUS LEADERS TO THE MEDIA.

城市也是**组织化**的中心。能源输入城市，为城市提供食品、热量、住房、消遣，并将不同的人联系起来，互相传播信息。这些人包括行政与管理人员，他们有的来自企业，有的来自政府，还有那些引导人们思想和观念的人们，如宗教领导人和大众媒体人等。

IN OTHER WORDS, CITY-DWELLERS MAKE DECISIONS THAT INFLUENCE THE FLOW OF ENERGY THROUGHOUT THE REST OF THE ECOSPHERE.

换句话说，城市居民作出的决定会影响到整个生态圈的能量流动。

TO SUSTAIN THE FLOW OF THINGS AND PEOPLE, TO BRING IN FOOD FOR MILLIONS, TO DISTRIBUTE IT TO COUNTLESS SHOPS AND RESTAURANTS, A CITY MUST HAVE A ***TRANSPORTATION SYSTEM.***

为了保证输送物品和居民行动的畅通无阻，保证为城市的数百万人运送食品，并同时分销到不计其数的商店和餐馆，城市一定要有一个**交通运输系统**。

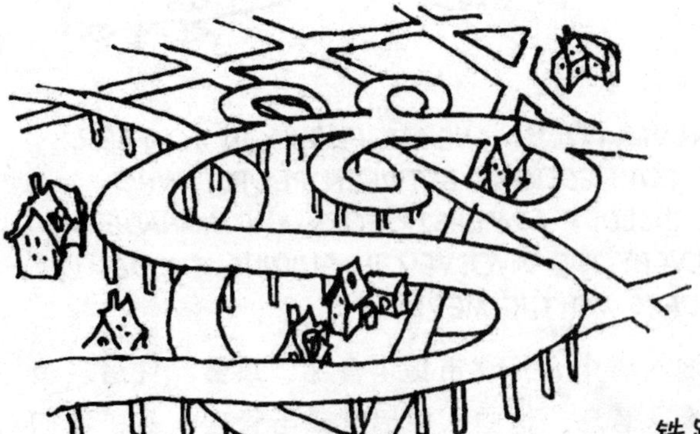

RAILROADS, RAIL YARDS, AIRPORTS, STREETS, HIGHWAYS, PARKING LOTS... IN SOME CITIES, LIKE LOS ANGELES, THE TRANSPORTATION SYSTEM COVERS MORE THAN HALF THE LAND AREA!

铁路、火车站、机场、街道、公路、停车场……在一些像洛杉矶这样的城市，交通运输系统的占地面积超过城市面积的一半。

WITHIN THE CITY, PEOPLE USE THREE TYPES OF TRANSPORTATION:

在城市里，人们使用三种交通工具：

IN THE U.S., MASS TRANSIT ACCOUNTS FOR ONLY 7% OF ALL PASSENGER TRAVEL, COMPARED WITH 15% IN THE FORMER WEST GERMANY, AND ALMOST 50% IN JAPAN. THE AMERICAN WAY IS STILL ONE CAR, ONE PERSON.

在美国，乘公共交通的人只占全部乘客的7%，而在前西德这一数字是15%，在日本达到了近50%。美国式的交通方式仍然是一人一车。

BESIDES THE TRANSPORTATION NETWORK, THERE HAVE TO BE OTHER SYSTEMS FOR DELIVERING SUCH THINGS AS ***WATER, ELECTRICITY, NATURAL GAS,*** ETC. A CITY NEEDS PIPELINES, AQUEDUCTS, RESERVOIRS, POWER PLANTS, AND WIRES.

除了交通运输网络外，城市还必须有其他的输送系统用来输送**水**、**电**、**天然气**等。城市需要输气管道、输水管道、水库、发电厂及电线。

THE CITY'S APPETITE FOR INPUT IS ENORMOUS. IN A TYPICAL DAY, THE AVERAGE U.S. CITY DWELLER USES AROUND ***100 GALLONS*** OF WATER, ***50 POUNDS*** OF FUEL (INCLUDING INDUSTRIAL USES), ***4 POUNDS*** OF FOOD, AND SOME CONSUMER GOODS.

城市的胃口是很庞大的。在一天之中，平均每个美国居民使用大约100**加仑***水、50**磅***燃料（包括工业用途）、4**磅**食物和一些消费品。

*1加仑（美制）≈0.004立方米，下同。

*1磅=0.4536千克，下同。

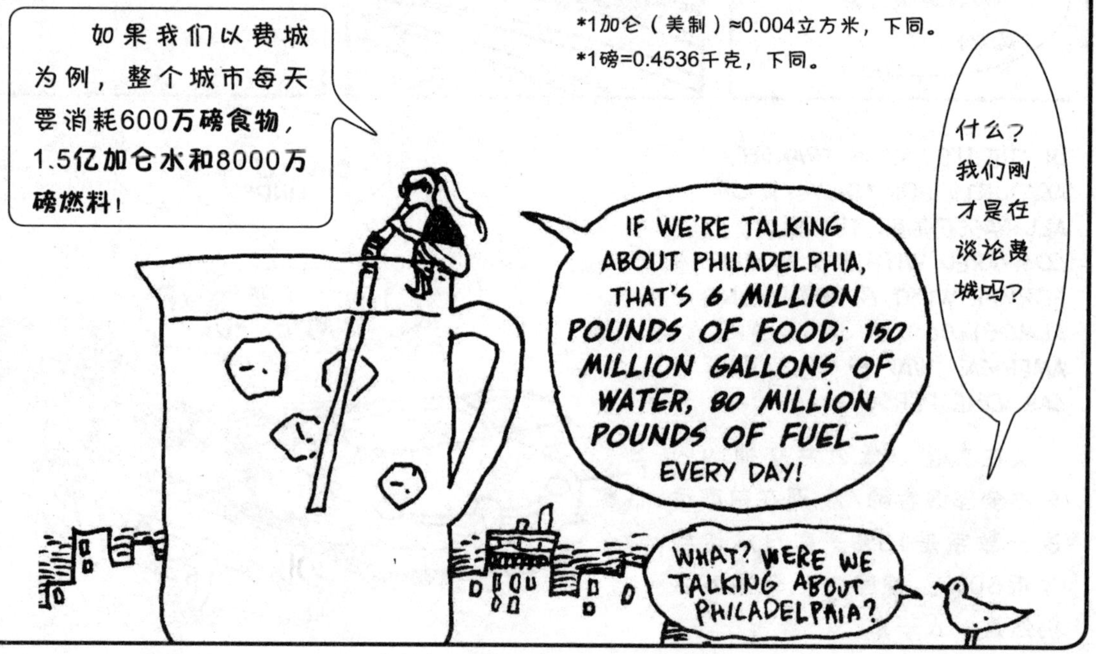

BRINGING IN GOODS AND ENERGY IS EASIER THAN GETTING RID OF THE OUTPUT PER PERSON, THAT'S ***4.5 POUNDS OF GARBAGE, A POUND OF HUMAN WASTE, 100 GALLONS OF WASTE WATER,*** AND ***A POUND OF AIR POLLUTION*** PER DAY.

将食品和能量运到城市里要比处理废物容易得多。每人每天会制造的废物包括**4.5磅垃圾**、**1磅排泄物**、**100加仑废水和1磅空气污染物**。

WHERE IS ALL THIS STUFF SUPPOSED TO GO? WE'VE OUTGROWN THE DAYS WHEN PEOPLE COULD JUST THROW SLOPS OUT THE WINDOW FOR PIGS TO EAT.

把这些垃圾运到哪里去呢？从前，我们把吃剩的饭菜扔到窗户外喂猪，这种日子已经一去不复返了。

FOR DISPOSAL PURPOSES, WASTE CAN BE DIVIDED INTO TWO KINDS: ***SOLID*** AND ***LIQUID*** (OR WASTEWATER).

从处理的角度划分，废物可分为两种：**固体废物**和**液体废物**（或称废水）。

让我们先讨论一下固体废物，俗称**垃圾**。

LET'S START WITH THE SOLID—OTHERWISE KNOWN AS

WHAT'S IN GARBAGE? ARCHAEOLOGIST/GARBOLOGIST WILLIAM RATHJE, WHO DIGS DUMPS, BREAKS IT DOWN LIKE THIS:

垃圾里都有什么？有一位名叫威廉·拉斯杰的考古学家和垃圾分析家曾挖开垃圾堆进行分析，结果如下：

40% PAPER AND CARDBOARD

40% 普通纸和硬纸板

20% CONSTRUCTION SCRAP AND YARD WASTE

20% 建筑垃圾和庭院垃圾

5-9% EACH
- FOOD WASTE 食物垃圾
- METAL 金属
- GLASS 玻璃
- PLASTIC 塑料

各占5%～9%

15% EVERYTHING ELSE: CLOTH, RUBBER, LEATHER, ETC.

15% 其他垃圾：衣物、橡胶、皮革等

THE FIRST ADVANCE OVER THE SLOP-AND-SCAVENGE SYSTEM CAME IN THE 1800s, WITH THE CREATION OF THE

SANITARY LANDFILL.

废物清除系统的首次伟大进步是在19世纪，即**垃圾填埋法**的发明。

THE IDEA HERE IS TO TRUCK TRASH OUTSIDE TOWN, DUMP IT, AND COVER EACH DAY'S DEPOSIT WITH A LAYER OF EARTH.

这种做法是：每天用卡车将废物运到城外，倒出来堆在一起，每一天的垃圾都盖上一层土。

BUT RAIN AND OTHER LIQUIDS CAN TRICKLE DOWN THROUGH THIS PILE AND CONTAMINATE THE GROUNDWATER.

但是雨水和其他液体会慢慢渗透到垃圾堆下面，污染地下水。

THE SOLUTION WAS TO BUILD ***LINED LANDFILLS*** WITH A LAYER OF CLAY OR PLASTIC TO PROTECT THE GROUNDWATER.

解决这一问题的方案是用黏土或塑料为**垃圾填埋场**衬砌保护层，来保护地下水。

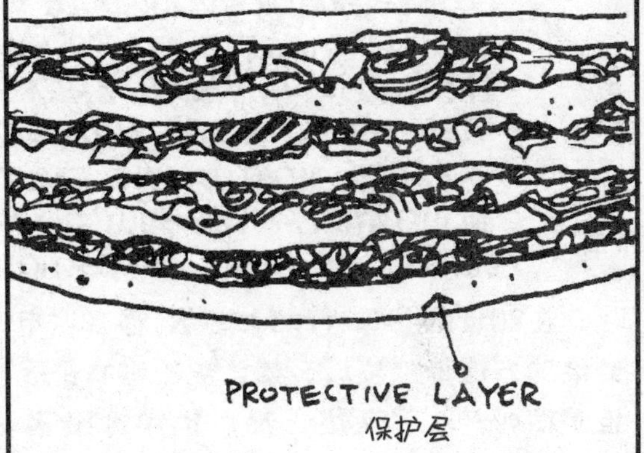

LANDFILLS ARE WIDELY USED... BUT THE GARBAGE IN THEIR DEPTHS BARELY DECOMPOSES, OR DOES SO ONLY SLOWLY... SO LANDFILLS ***ONLY GET BIGGER***. AS OLD ONES FILL UP, NEW SITES MUST BE FOUND, OFTEN VERY FAR FROM THE CITIES THEY SERVE.

垃圾填埋场被广泛应用……但是处于地下深层的垃圾几乎不降解，或者降解的速度非常慢……因此垃圾堆只会**越积越厚**。当旧的填埋场被填满后，就必须找到新的填埋场。通常会建在离市区很远的地方。

IN THE 1940s AND 1950s, MANY CITIES TRIED TO SHRINK GARBAGE BY BURNING IT IN

INCINERATORS.

INCINERATORS DO REDUCE GARBAGE VOLUME, BUT THEY ALSO PRODUCE **ASH, HEAT, CHEMICAL POLLUTANTS,** AND **BAD SMELLS.** MOST OF THE EARLY INCINERATORS WERE SHUT DOWN BECAUSE OF PUBLIC OPPOSITION.

在20世纪40年代和50年代，很多城市试着用**焚化炉**焚烧的方法消除垃圾。焚化炉的确减少了垃圾的数量，但是同时产生了**灰烬、热、化学污染物**以及**难闻的气味**。大部分早期的焚化炉因公众的反对而关闭。

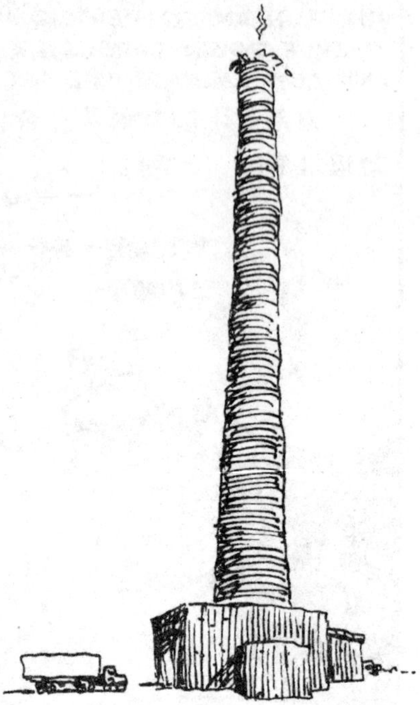

IN THE 1970s AND 80s, INTEREST REVIVED IN HIGH-TECH ***WASTE-TO-ENERGY*** INCINERATORS, WHERE HEAT FROM BURNING GARBAGE WAS SUPPOSED TO GENERATE ELECTRICITY. THESE STILL HAD EMISSIONS PROBLEMS: TO BURN CLEANLY, THEY WOULD HAVE TO RUN CONTINUOUSLY, REQUIRING A STEADY STREAM OF PRE-SHREDDED GARBAGE, AND STILL NEED EXPENSIVE POLLUTION-CONTROL DEVICES.

20世纪70年代和80年代，人们对**废物转化为能源**的高科技焚化炉再次产生兴趣。用这种焚化炉可以将焚烧垃圾产生的热能转化成电能。但这种方法仍然有排放废物的问题。为了烧得干净，焚化炉必须要连续作业，同时也要求连续不断加入提前粉碎好的垃圾，而且还需要昂贵的控制污染的装置。

BESIDES, ENVIRONMENTALISTS KEPT POINTING OUT, IF GARBAGE PRODUCTION COULD ONLY BE CUT DOWN, INCINERATORS WOULDN'T BE NEEDED IN THE FIRST PLACE.

除此之外，环保主义者不断指出，如果少制造垃圾，那就不需要垃圾焚化炉了。

AND HOW DO YOU CUT GARBAGE PRODUCTION?

怎样才能减少垃圾呢？

RECYCLING

AS YOU CAN SEE FROM RATHJE'S LIST THREE PAGES BACK, A LOT OF "WASTE" CAN ACTUALLY BE REUSED AGAIN AND AGAIN. ALL IT TAKES IS SORTING THE STUFF OUT AHEAD OF TIME AND SENDING IT TO A RECYCLING PLANT—OR MAKING YOUR OWN GARDEN COMPOST—INSTEAD OF JUST THROWING IT "AWAY."

垃圾回收利用

正如你从威廉·拉斯杰在列出的垃圾分类目录（第214页）上所看到的，事实上许多“垃圾”能被一次一次地再利用。我们所要做的仅仅是把东西提前分类，再送到回收厂，或者把垃圾用来做花园里的肥料，而不是把它扔掉而已。

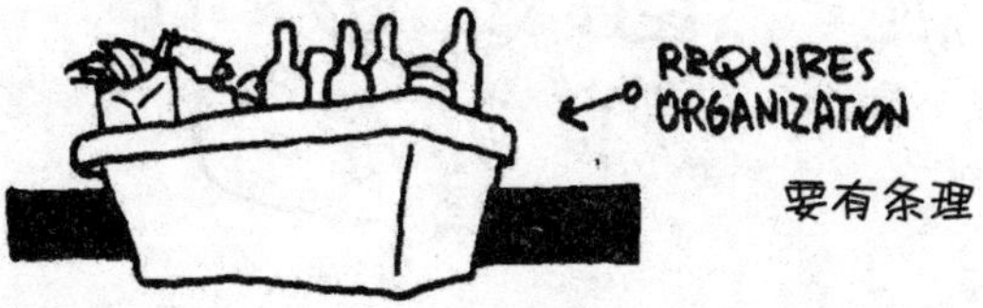

FOR RECYCLING TO WORK, INDUSTRY HAS TO HAVE THE CAPACITY TO USE THE SECOND-HAND MATERIAL. ABOUT ***60%*** OF ALUMINUM, FOR EXAMPLE, NOW GOES BACK FOR RESMELTING, BUT AMERICAN PAPER MILLS STILL RECYCLE LITTLE. INSTEAD, RECYCLED PAPER GOES TO ***EUROPE*** AND ***JAPAN***, WHERE TREES ARE MORE EXPENSIVE.

要想实施循环再利用，工业必须有能力利用回收材料。例如，现在美国大约有60%的铝被回收再熔炼，但美国的造纸厂却很少回收利用纸。**欧洲**和**日本**的木材很昂贵，所以常使用再生纸。

THE OTHER OBVIOUS WAY TO MAKE LESS GARBAGE IS TO USE ***LESS DISPOSABLE STUFF***. IN THE JARGON, THIS IS CALLED

SOURCE REDUCTION.

另外一种明显减少垃圾的方法就是少使用**一次性物品**。用行话来说，就是实施**“源头减量”**。

THE MOST ENERGY-EFFICIENT WAY TO MANAGE SOLID WASTE: JUST DON'T MAKE ANY!!

最节能的控制固体垃圾的方法是：不制造垃圾！！

SOURCE REDUCTION IS ABOUT ***LIFESPAN:*** WE NEED TO USE THINGS THAT LAST LONGER, FROM SOCKS TO HOUSES. INSTEAD OF DEBATING THE RELATIVE MERITS OF ***PLASTIC*** OR ***PAPER*** BAGS AND CUPS, TAKE YOUR OWN ***CLOTH BAG*** TO THE STORE... AND USE A CERAMIC CUP!

源头减量牵涉到产品的使用**寿命**：我们应该选择耐用、使用寿命较长的产品，无论是袜子还是房子都是如此。去商场时自己带个**布袋**，而不是争论**塑料**袋或**纸**袋、**塑料**杯或**纸**杯的优缺点……再就是使用陶瓷杯子！

SOURCE REDUCTION AND RECYCLING PROGRAMS HAVE BEEN BIG SUCCESSES WHERE THEY'VE BEEN TRIED. IN SEATTLE, FOR EXAMPLE, PER CAPITA GARBAGE PRODUCTION FELL BY ABOUT **65%** BETWEEN 1983 AND 1993, AND FEW PEOPLE ARE TALKING ABOUT INCINERATORS ANY MORE.

那些实验了源头减量和废物回收项目的地方都取得了很大的成功。例如在西雅图，人均垃圾制造量在1983～1993年下降了大约65%，很少有人再谈论焚化炉了。

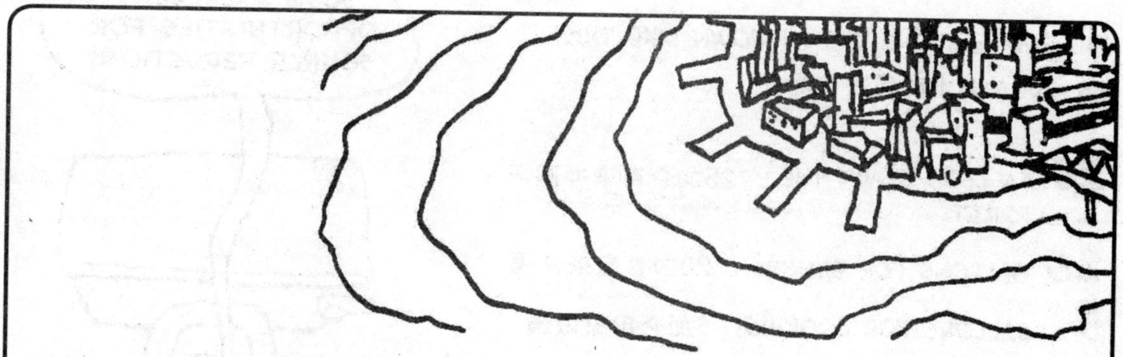

FOR WATER, THE PROBLEMS ARE A LITTLE DIFFERENT: WATER WON'T JUST SIT THERE LIKE GARBAGE. IT RUNS AWAY, SPREADING WHATEVER IS IN IT ALL OVER THE PLACE.

对于水来说，问题就有点不同了。水不会像垃圾一样静静地待在那里。它到处流淌，漫过所有能及的地方。

THE FIRST PROBLEM WITH WATER IS GETTING ***ENOUGH OF IT:*** AMERICANS, WHOSE CITIES HAVE GOOD PLUMBING, USE ABOUT ***100 GALLONS PER PERSON*** EVERY DAY. IN A CITY OF MILLIONS THAT ADDS UP, AND NOW NEW YORK COMPETES WITH PHILADELPHIA FOR WATER, WHILE LOS ANGELES FIGHTS WITH THE ENTIRE STATE OF ARIZONA OVER THE SOUTHWEST'S MEAGER SUPPLY.

关于水，首先要解决的问题就是获取**足够多的水**。美国城市有先进的管道系统，每天**人均要用掉大约100加仑水**。在一个有数百万人口的城市里，用水量还会更多。现在纽约和费城在争夺水资源，洛杉矶和整个亚利桑那州也在争夺西南地区本来就紧缺的水资源。

THE 100 GALLONS BREAKS DOWN LIKE THIS:

100加仑水的用途分为：

25 GALLONS DOWN THE TOILET　25加仑用来冲厕所

20 GALLONS FOR BATHING　20加仑用来洗澡

5 GALLONS FOR COOKING　5加仑用来做饭

15 GALLONS FOR DISHWASHING　15加仑用来洗碗筷

20 GALLONS FOR WASHING CLOTHES　20加仑用来洗衣服

15 GALLONS FOR MISCELLANEOUS HOUSE & GARDEN USES　15加仑用于家中和花园的各种其他需要

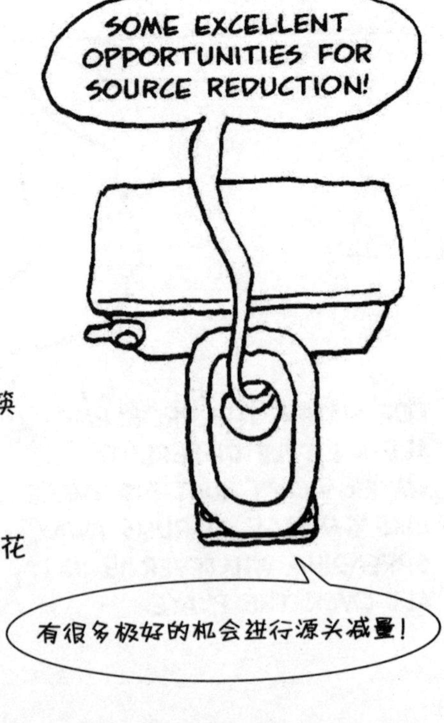

IN WATER-POOR CALIFORNIA, STRICT REGULATIONS HAVE CUT PER CAPITA WATER CONSUMPTION IN HALF. SOME OF THE WAYS:

在水资源贫乏的加利福尼亚州，通过严格的管理制度，人们已经将人均用水量减少了一半。其中一些做法是：

* USE LOW-FLOW SHOWER HEADS
  * 使用慢水流淋浴喷头
* USE TOILETS WITH SMALLER TANKS—OR PUT A BRICK IN YOUR OLD TANK
  * 使用水箱较小的马桶或者在旧水箱里放块砖
* DON'T FLUSH FOR EVERYTHING
  * 不要什么东西都冲洗
* WATER THE GARDEN BY DRIP INSTEAD OF SPRINKLER (LESS EVAPORATION!)
  * 用滴灌的方式浇花园而不用喷洒式（减少水分蒸发）浇灌花园
* TURN OFF FAUCET WHILE BRUSHING TEETH.
  * 刷牙的时候关掉水龙头

MEANWHILE, AT THE OTHER END OF THE PIPE:

与此同时，在管道的另一端：

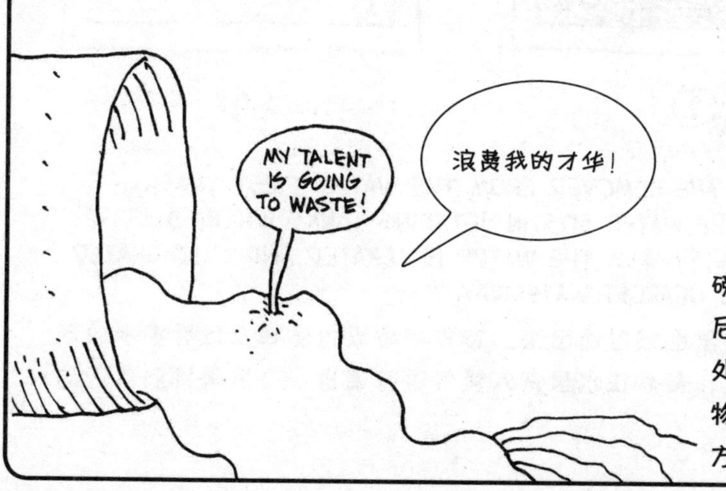

A LARGE CITY PUTS MILLIONS OF POUNDS OF FECES AND URINE INTO THE WATER EVERY DAY, AND DISCHARGES THEM AT A SINGLE POINT. UNTREATED WASTE WATER IS THEREFORE RICH IN ORGANIC MATTER THAT WAS ONCE (IN SOME PLACES) APPRECIATED FOR ITS FERTILIZING POWER.

一个大城市每天把数百万磅的粪便和尿排放在水中，然后把它们排到某处。因此未经处理的废水里富含丰富的有机物，这些有机物（在一些地方）曾是非常好的肥料。

DUMPED STRAIGHT INTO RIVERS AND OCEANS, HOWEVER, THIS LOAD OF ORGANIC MATTER OVERWHELMS THE ECOSYSTEM: BACTERIA AND ALGAE FLOURISH IN IT, AND AFTER A CERTAIN POINT, FISH MAY ASPHYXIATE.

然而，废水被直接排到河里和海里，丰富的有机物给生态系统带来沉重的负担：细菌和藻类在里面滋生，到一定程度时鱼就会窒息。

YET UNTREATED DUMPING IS WHAT MOST MAJOR CITIES DID UNTIL RECENTLY, WHEN THEY BEGAN BUILDING MODERN ***WASTEWATER TREATMENT PLANTS.***

然而，在过去很长一段时间里，大部分大城市都是将未经处理的污水直接排掉，直到近些年才开始建造现代化的**污水处理厂**。

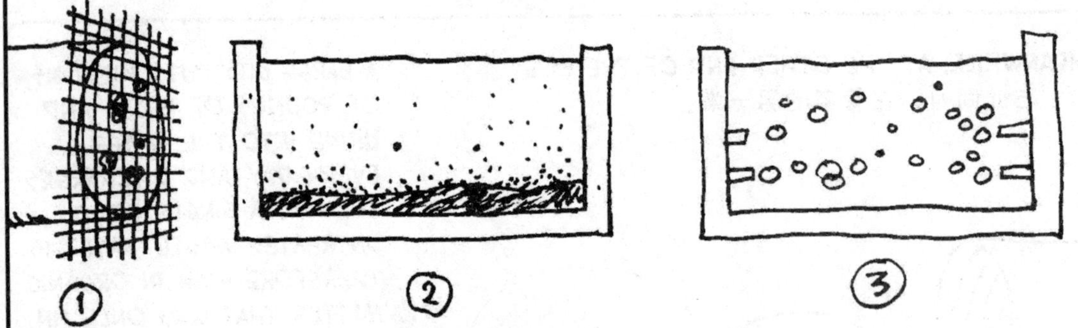

IN THESE FACILITIES, SOLIDS ARE REMOVED FROM THE WATER: FIRST TRASH IS SCREENED OUT, AND THEN THE WATER SITS IN SETTLING TANKS WHERE SLUDGE CAN SETTLE TO THE BOTTOM. FINALLY, THE WATER IS AERATED AND CHLORINATED BEFORE DISCHARGE INTO THE NEAREST WATERWAY.

通过这些设施，固体从水里面被过滤出来。首先将垃圾过滤掉，然后把水放到沉淀池里，让沉淀物沉到池底，最后往水里充入氯气进行消毒，之后再排到最近的水道里。

THE SLUDGE IS COLLECTED--AND THEN WHAT?

沉淀物被收集起来——之后呢?

IN THE PAST, SLUDGE WAS DUMPED OR INCINERATED, BUT MORE RECENTLY, AS THE U.S. HAS BANNED POURING POISONS DOWN THE DRAIN, SLUDGE HAS BECOME CLEAN ENOUGH TO BE TURNED INTO FERTILIZER PELLETS AND SOLD.

过去，或者将沉淀物直接扔弃，或者焚烧。但是最近由于美国已经禁止将有毒物质倒入下水道，因此沉淀物被转化成非常干净的颗粒状农业肥料进行出售。

THESE PURIFICATION PLANTS DEPEND ON GOOD CITYWIDE PLUMBING TO GATHER THE RAW SEWAGE FOR TREATMENT, BUT THERE ARE MANY HUGE CITIES WITHOUT GOOD SEWERS. IN BOMBAY, CAIRO, AND MEXICO CITY, WHERE MILLIONS HAVE ONLY THE GUTTER FOR A TOILET, WASTE-DISPOSAL PROBLEMS ARE BEYOND THE SCOPE OF THIS BOOK—AND ALMOST BEYOND BELIEF.

这些净化厂依赖覆盖全市的先进管道系统去收集污水进行处理，但世界上很多大城市没有这种先进的污水管。在孟买、开罗和墨西哥城，数百万人竟然将排水沟当厕所。这些地方的废水处理问题超出了这本书的讨论范围——几乎也超出了我们的想象。

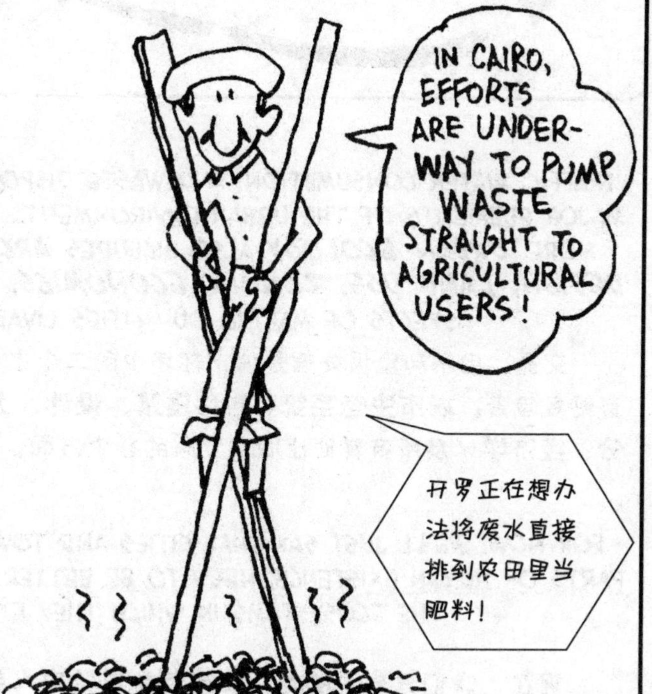

NOR DO WE HAVE ROOM TO DISCUSS THE CRUCIAL ROLE OF WETLANDS AS NATURAL WATER PURIFICATION SYSTEMS... BUT HERE'S THE GENERAL IDEA:

我们也没有更多的篇幅去讨论湿地作为天然的水资源净化系统所扮演的重要角色……基本概念如下：

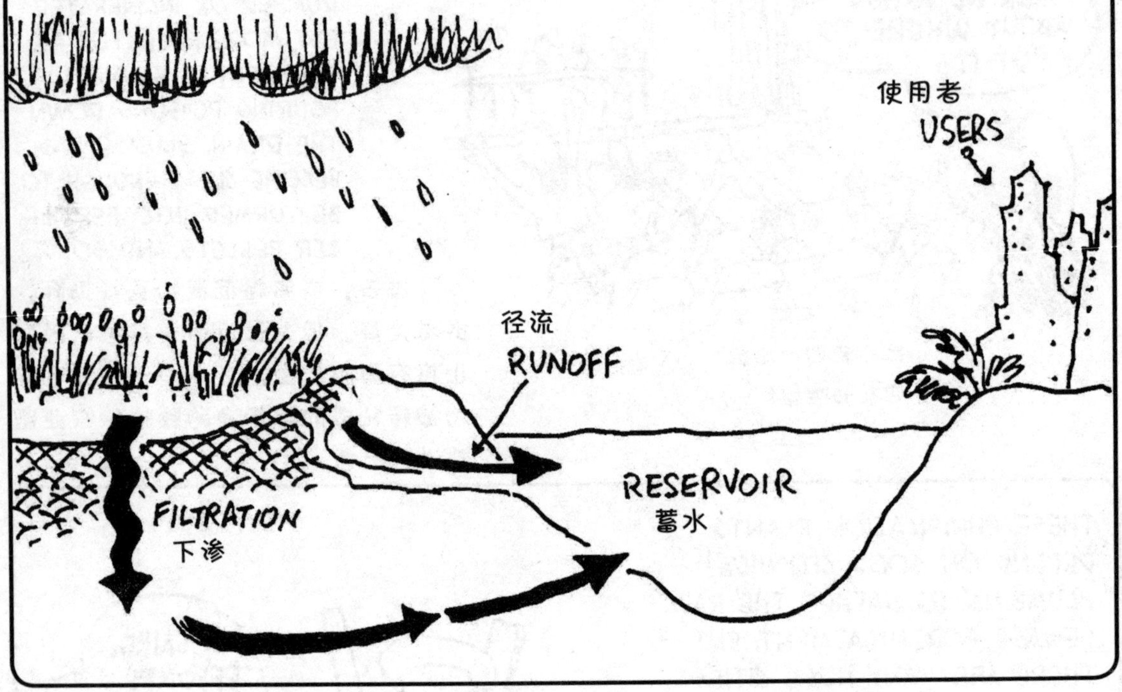

TRAFFIC, WATER CONSUMPTION, AND WASTE DISPOSAL ARE THREE MAJOR ELEMENTS OF THE URBAN ENVIRONMENT... AND THERE ARE MORE. ***URBAN ECOLOGY*** ALSO INCLUDES ***ARCHITECTURE, DESIGN, LAND USE, ZONING, ECONOMICS***, AND ALL OTHER ASPECTS OF MAKING OUR CITIES LIVABLE.

交通、用水和垃圾处理是城市环境中的三个主要因素……当然还有更多因素。**城市生态系统**也包括**建筑**、**设计**、**土地利用**、**区域划分**、**经济学**以及所有其他让城市宜居的各个方面。

FOR NOW, WE'LL JUST SAY THAT CITIES AND TOWNS, LIKE OTHER PARTS OF HUMAN EXISTENCE, NEED TO BE BETTER INTEGRATED WITH THE ECOSYSTEMS IN WHICH THEY EXIST.

现在，我们只想说的是，城市与乡镇就像人类生活的其他方方面面，需要更好地与我们生存的这个生态系统融为一体。

MAYBE THERE IS HOPE FOR THE URBAN PIG!
城市中的猪可能有希望了！

·CHAPTER 13·

POLLUTION

EVER SINCE PEOPLE BEGAN CONSUMING, WE'VE THROWN THE LEFTOVERS AWAY... BUT WHERE IS "AWAY?"

◆第十三章◆

污染

自从人类开始消费，我们就一直在扔弃废物……但是都“扔到”哪里了呢？

WHEN POPULATION WAS SMALL AND WASTES WERE ALL ORGANIC, THIS WAS LESS OF AN ISSUE. THE WORLD WAS LARGE, AND OUR LEAVINGS WERE DILUTED OR DEGRADED UNTIL THEY ESSENTIALLY DISAPPEARED.

当人口数量不多且所有的废弃物都是有机物时，上述问题并不严重。而且我们生活的世界很大，废弃物都会自然分解直到最后彻底消失。

BUT THIS CHANGED WITH THE INDUSTRIAL REVOLUTION, WHEN POLLUTION BECAME A

BIG PROBLEM.

但是随着工业革命的到来，这一切都改变了，环境污染便成为一个**大问题**。

THERE WERE AT LEAST THREE REASONS:

最少有三个原因导致了这个问题：

1. ***INDUSTRIAL PROCESSES:*** MORE SMOKE WENT INTO THE AIR, AND MORE WASTE FLOWED INTO THE WATER.

工业生产过程：越来越多的工业气体被排放到空气中，越来越多的废弃物流入水中。

2. ***POPULATION GROWTH:*** MORE STUFF TO THROW AWAY, AND LESS PLACE TO THROW IT.

人口的增加：需要扔的东西越来越多，而扔废物的地方却越来越少。

3. ***CHEMISTRY:*** BEGINNING IN 1828, WITH THE SYNTHESIS OF UREA, CHEMISTS HAVE CONSTRUCTED OR DISCOVERED SOME ***7 MILLION*** NEW SUBSTANCES: DYES, EXPLOSIVES, PLASTICS, DETERGENTS, SOLVENTS, ETC. IT'S NOT JUST THAT WE DUMP MORE INTO THE ENVIRONMENT: THE NATURE OF WHAT WE DUMP HAS ALSO CHANGED.

化学品：从1828年化学物质尿素的首次合成到现在，化学家又合成了大约700万种新的化学物质，有染料、炸药、塑料、洗涤剂、溶剂等。我们不仅把更多的废物扔到了环境中，而且这些废弃物的本质特点也已经改变了。

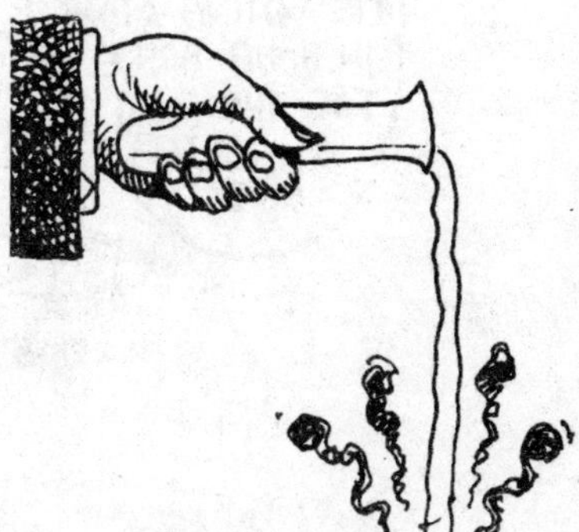

THE FIRST INKLING THAT RELEASING NEW SUBSTANCES CAN HAVE UNEXPECTED RESULTS CAME IN THE 1950s. ***NUCLEAR BOMBS***, EXPLODED ABOVEGROUND IN REMOTE LOCATIONS, CREATED ***RADIOACTIVE FALLOUT*** THAT WAS DETECTED IN AIR, RAIN, FOOD, SOIL, AND WATER ALL OVER THE WORLD.

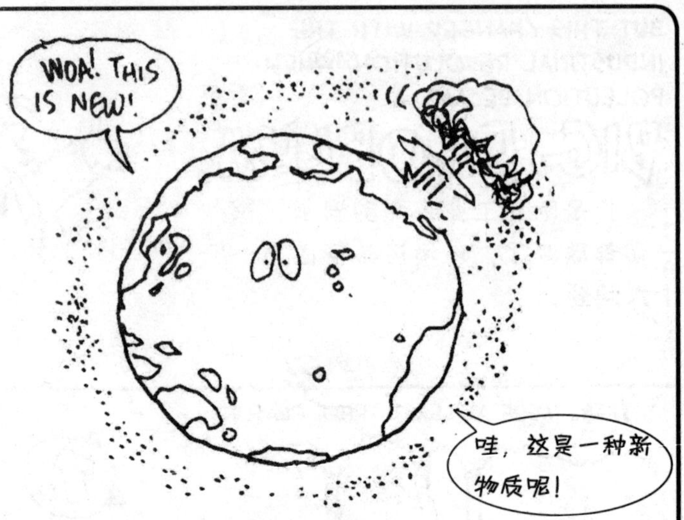

20世纪50年代，人类第一次感觉到释放新的物质会带来意想不到的后果。在远离人烟的地面上空进行的**核弹**爆炸实验产生了**放射性沉降物**，世界各地的空气中、雨水中、食物里、土壤里、水里，都检测到了这类物质。

IN 1954, ***STRONTIUM-90,*** A RADIOACTIVE ELEMENT SIMILAR TO CALCIUM, WAS FOUND IN COWS' MILK AND CHILDREN'S BONES EVERYWHERE ON EARTH.

1954年，在全世界许多地区的牛奶和小孩的骨骼中发现了**锶-90**，这是一种类似于钙元素的放射性元素。

GOVERNMENTS ADDRESSED THIS PROBLEM BY AGREEING TO TEST WEAPONS DEEP UNDERGROUND, WHERE NO RADIOACTIVITY COULD REACH THE ATMOSPHERE.

为了解决这个问题，政府同意在很深的地下进行武器试验，这样放射性物质就不会释放到大气里了。

THE NEXT WAKE-UP CALL CAME FROM THE PESTICIDE

D.D.T.

另一个警钟来自于杀虫剂DDT的使用。

SPRAYED ON CROPS TO KIL INSECTS, TRACES OF DDT TRICKLED INTO PONDS AND STREAMS...

DDT被用来喷洒在农作物上除掉害虫，而其残留物则会慢慢地渗到池塘和溪流中。

BUT DDT HAS TWO NASTY PROPERTIES: IT'S LONG-LIVED, AND IT ACCUMULATES IN FATTY TISSUE.

但DDT有两个非常有害的特性：一是存留时间长，二是能在脂肪组织里大量富集。

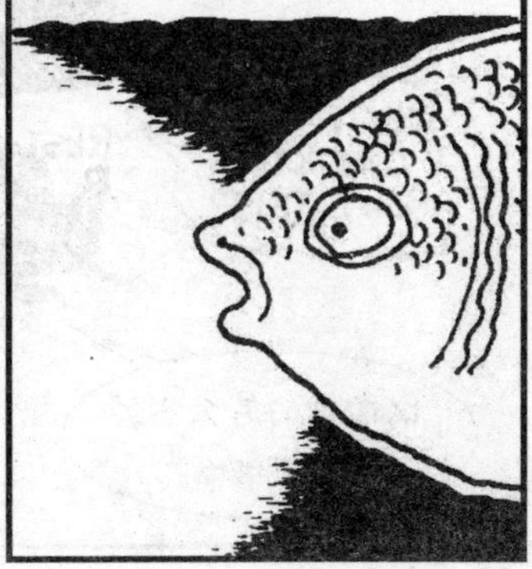

这意味着生物体内都会有DDT含量，处于食物链越高层级的生物体，DDT含量越高。如：

BY THE 1960s, PELICAN POPULATIONS HAD PLUMMETED, AND BALD EAGLES HAD NEARLY DISAPPEARED.

到了20世纪60年代，鹈鹕的数量急剧减少，白头海雕几乎完全消失了。

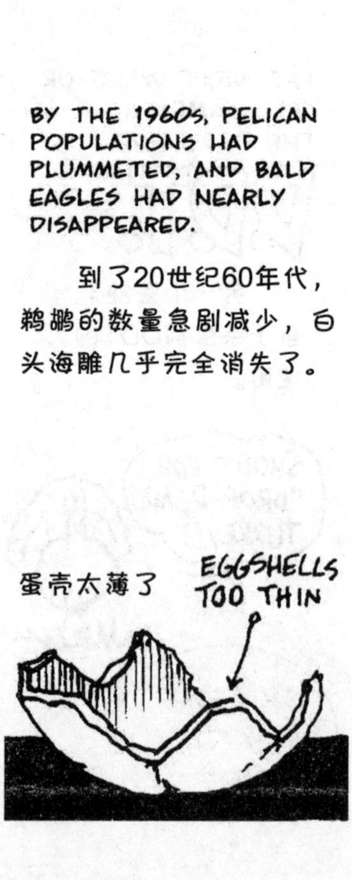

IN 1962, SCIENCE WRITER

RACHEL CARSON

MADE AMERICA TREMBLE WITH HER BOOK ***THE SILENT SPRING.***

1962年，科普作家**雷切尔·卡逊出版的《寂静的春天》**，震惊了所有的美国人。

CARSON WARNED THAT IF DDT AND OTHER PERSISTENT CHEMICALS WENT UNCHECKED, THEY WOULD SPREAD THROUGHOUT THE ENVIRONMENT, KILLING WILDLIFE AND RAISING CANCER RATES AMONG HUMANS. ***CHANGE COURSE,*** SHE SAID, OR RISK ***POISONING THE EARTH.***

卡逊警告说，如果滥用DDT和其他不易分解的化学制品的话，这些化学制品将会蔓延到整个环境中，会杀死野生动物并且会提高人类患癌症的概率。她说道，要么**改变现状**，要么**毒害地球**。

OVER THE NEXT FEW YEARS, ENVIRONMENTAL QUALITY PLUNGED... THE HUDSON RIVER DIED... THE GREAT LAKES GASPED... THE OIL-CHOKED CUYAHOGA RIVER ACTUALLY ***CAUGHT FIRE...*** (!)

在接下来的几年中，环境质量急剧下降……哈德逊河死了……五大湖奄奄一息了……而充满油污的凯霍加河真的已经**着火了**……（！）

GRADUALLY IT DAWNED ON PEOPLE THAT YOU CAN'T THROW THINGS AWAY, BECAUSE THERE IS NO "AWAY"—AND SO BEGAN THE BIGGEST SOCIAL MOVEMENT IN U.S. HISTORY:

ENVIRONMENTALISM.

于是人类逐渐明白,不能随意把东西扔弃,因为没有可以扔东西的地方了。于是，美国历史上规模最大的社会运动开始了，即**环境保护主义运动**。

绿色最美!

IN 1972, THE UNITED STATES PASSED THE MOST STRINGENT POLLUTION-CONTROL REGULATIONS IN THE WORLD, AND CREATED AN

environmental protection agency

TO OVERSEE COMPLIANCE.

1972年，美国政府通过了世界上最严厉的控制环境污染的法规，并且成立了**环境保护局**以便监督遵守环境保护情况。

尼克松

THE E.P.A. SET LIMITS ON INDUSTRIAL AND AUTOMOTIVE EMISSIONS OF VARIOUS KINDS:

环境保护局对各种工业排放物和汽车排放物进行了限制。

TOXIC

SUBSTANCES ARE THOSE THAT ARE JUST PLAIN POISON. AT LOW DOSES, THEY ARE FATAL TO HUMANS.

有毒物质是指那些有明显毒性的物质。这种物质只要很小的剂量就会对人类造成致命的伤害。

HAZARDOUS

WASTES ARE THOSE THAT MAY BURN, DISSOLVE THINGS, EXPLODE, IRRITATE, OR CAUSE ALLERGIC REACTIONS.

危险废品是指那些可燃烧、有腐蚀性、易爆、有刺激性或能造成过敏反应的物质。

CARCINOGENS

INCREASE THE RISK OF CANCER. THESE INCLUDE NOT ONLY MANY CHEMICALS BUT ALSO ANYTHING THAT EMITS RADIATION.

致癌物增加了人们患癌症的风险。这些物质不仅包括许多化学品，也包括任何具有放射性的物质。

THE ALLOWABLE DISCHARGE LEVELS OF THESE CONTROLLED SUBSTANCES ARE BASED ON ***RISK-BENEFIT*** OR ***COST-BENEFIT ANALYSIS***, IN WHICH THE RISKS (OR COSTS) OF POLLUTION ARE BALANCED AGAINST THE BENEFITS OF USING THE CHEMICAL.

对这些限制物的允许排放水平是基于**风险-收益**或**成本-收益分析**考虑的。在这种分析里，使用化学品污染环境的风险（或成本）必须要与使用该化学品的收益相平衡。

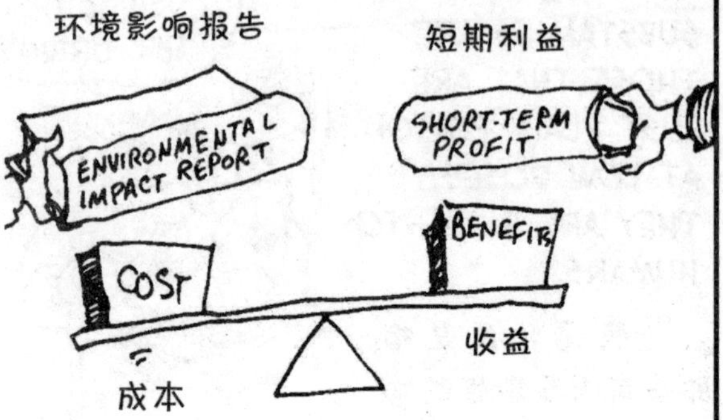

THIS MAY SOUND REASONABLE. EVERYTHING IN LIFE IS A TRADE-OFF, RIGHT?

这听起来似乎很合理。生活中的一切都是有利有弊的，不是吗？

THE PROBLEM IS THIS: BENEFITS ARE OFTEN EASY TO SEE, BUT COSTS CAN BE HARD TO ASSESS, HIDDEN, OR ONLY APPARENT LATER ON.

然而问题是：收益通常是显而易见的，而成本却很难去估计，因为成本往往是隐性的，或者以后才会显现出来。

PESTICIDES, FOR EXAMPLE, RAISE CROP YIELDS, LOWERING FOOD PRICES FOR EVERYONE...

比如说，农药提高了粮食的产量，从而降低了粮食的价格……

BUT PESTICIDES ARE POISONOUS. HOW DO YOU BALANCE THE BENEFIT AGAINST THE HEALTH DAMAGE TO FARMWORKERS?

但是农药是有毒的。你怎么来权衡这种收益和它给农民的健康带来的危害呢？

AND WHAT DO YOU DO WHEN YOU LATER DISCOVER THAT PESTICIDE TOXINS ARE SHOWING UP IN GROUNDWATER? OR THAT INSECTS, MASTER **r**-STRATEGISTS THAT THEY ARE, BECOME ***RESISTANT*** TO ONE PESTICIDE AFTER ANOTHER?

那么，当我们有一天在地下水中发现了农药中的有毒物质时，或者当我们发现掌握了繁殖策略的害虫对一种又一种农药都产生了**抗药性**时，我们该怎么办呢？

DESPITE SUCH PROBLEMS, THERE HAVE BEEN MANY ENVIRONMENTAL SUCCESS STORIES SINCE 1972. DDT WAS BANNED OUTRIGHT (FOR USE IN THE USA—AMERICAN COMPANIES CAN AND DO STILL SELL IT ABROAD!), AND BIRD POPULATIONS HAVE REBOUNDED SOMEWHAT.

尽管产生了这些问题，但自1972年以来，成功保护环境的例子也有很多。DDT被彻底禁止使用了（只在美国禁止使用——但美国的公司可以向国外出售，而且依然在出售），鸟类的数量有所增长。

BEFORE 1972, HAZARDOUS CHEMICALS WENT DOWN THE DRAIN... NOW THEY GO INTO ***LICENSED, MONITORED, DOUBLE-LINED DISPOSAL FACILITIES*** OR ARE INCINERATED UNDER STRINGENT EMISSION CONTROLS.

1972年以前，危险的化学品都被排放到了下水道里……而现在，这些危险的化学品进入**被批准的、受到监控的、双层隔离的**废品处理设备中加以处理，或者在严格的排放控制下被焚化。

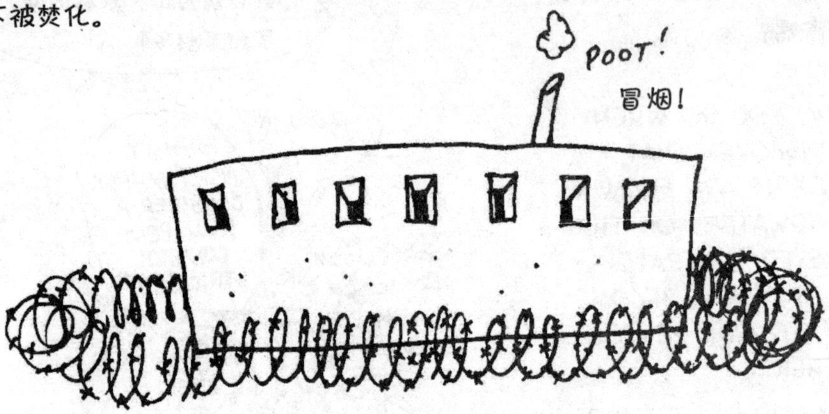

BUT ENFORCEMENT IS IMPERFECT... STANDARDS IN MANY COUNTRIES ARE WEAK... AND BEFORE 1972 WE HAD ALREADY PUMPED BILLIONS OF TONS OF CHEMICALS INTO THE ENVIRONMENT THAT WE CAN'T EASILY TAKE BACK.

但是实施起来并不理想……许多国家制定的标准都不十分严格……而且在1972年以前，我们已经向环境中排放了数十亿吨的化学品，这些排放物很难被收回。

WHILE TOXINS WERE A HIDDEN PROBLEM THAT SNEAKED UP ON US, ***AIR POLLUTION*** WAS RIGHT UNDER OUR NOSES. THE COMBUSTION OF FOSSIL FUELS IN FACTORIES AND CARS PRODUCES A HOST OF NOXIOUS STUFF: ***NITROGEN OXIDES, SULFUR DIOXIDE, CARBON MONOXIDE...*** AND MANY INDUSTRIAL CHEMICALS ARE GASEOUS AS WELL. THE IMMEDIATE AND OBVIOUS RESULT WAS

如果说有毒物质是一个潜伏的问题，在悄悄地威胁着我们的健康，那么，**空气污染**就是我们的鼻子能够嗅到的问题了。工厂和汽车燃烧的化石燃烧排放出了大量的有毒物质：**氮氧化物**、**二氧化硫**、**一氧化碳**……加上许多化工原料本身也是气体，这样导致的最直接最明显的结果就是产生了大量的**烟雾**。

IN THE 1970S, AIR POLLUTION WAS TREATED AS A LOCAL PROBLEM: IT HAPPENED MAINLY IN CITIES. THE ***CLEAN AIR ACT*** IMPOSED LIMITS ON VARIOUS EMISSIONS, SO THAT CARS NOW COME WITH POLLUTION-CUTTING ***CATALYTIC CONVERTORS...*** WHILE POLLUTING INDUSTRIES BOUGHT SCRUBBERS, BAGHOUSE FILTERS, ELECTROSTATIC PRECIPITATORS, AND CYCLONE SEPARATORS TO CLEAN UP THEIR FACTORY SMOKE.

20世纪70年代，空气污染被当作一个地方性问题来对待：这种问题主要产生在城市里。**清洁空气法**对许多排放物进行了限制，所以现在的汽车都装有具有减排功能的**催化转换器**……一些制造空气污染的工业购置了洗涤器、袋式隔滤装置、静电除尘器和旋风除尘器来清除工厂产生的烟雾。

RESULT: ANOTHER (LOCAL) SUCCESS. A NEW CAR NOW EMITS ABOUT **2%** THE POLLUTION OF A 1970 MODEL... AMERICAN FACTORIES ARE CLEANER, AND MOST AMERICAN CITIES' AIR IS MORE BREATHABLE NOW THAN IT WAS 20 YEARS AGO.

结果又是一个（具有地方特色的）成功案例。新款汽车尾气的排放量只是1970年款汽车尾气排放量的2%……美国的工厂比以前更干净了，美国的大多数城市的空气质量也比20年前更好了。

BUT ATMOSPHERIC CIRCULATION IS WORLDWIDE, AND ATMOSPHERIC CHEMISTRY IS COMPLEX. THE ECOSYSTEM MOVED THOSE GASEOUS POLLUTANTS AROUND AND TRANSFORMED THEM FROM LOCAL SMOG INTO SOME REGIONAL AND EVEN GLOBAL SURPRISES.

但是大气循环是世界范围的，而且大气化学也是很复杂的。生态系统使气体污染物到处流动，而且把小范围的烟雾转化成地区性的甚至是全球性的污染。

ACID RAIN

IS CAUSED WHEN ***SULFUR DIOXIDE*** (SO_2) AND ***NITROGEN OXIDES*** (N_2O, NO, AND NO_2) IN THE AIR REACT WITH OTHER GASES TO FORM ***SULFURIC ACID*** (H_2SO_4) AND ***NITRIC ACID*** (HNO_3), TWO OF THE STRONGEST ACIDS IN NATURE. DISSOLVING IN RAINWATER, THESE ACIDS FALL TO EARTH.

酸雨的形成是这样的：当空气中的**二氧化硫**（SO_2）和**氮氧化物**（N_2O、NO和NO_2）以及其他气体相互作用便形成了**硫酸**（H_2SO_4）和**硝酸**（HNO_3），这是自然界中两种最强的酸。这两种酸溶解在雨水中然后落到地面上，便形成酸雨。

WHEN ACID RAIN HITS THE GROUND, IT RELEASES METAL IONS FROM THE SOIL: ***ALUMINUM, CADMIUM, MERCURY,*** AND ***LEAD,*** WHICH LEACH INTO THE AQUIFER, POISONING FISH—AND ANIMALS (SUCH AS PEOPLE!) THAT EAT FISH.

当酸雨落到地面上时，它会释放出土壤中的一些金属离子：如**铝离子**、**镉离子**、**汞离子**和**铅离子**，金属离子慢慢渗入地下含水层中，毒害鱼类，也毒害食用这些鱼的动物（如人类！）。

EVEN WITHOUT THE METALS, FISH CAN'T SURVIVE IN A STRONGLY ACID ENVIRONMENT... AND TODAY, MANY NORTHERN LAKES ARE ACIDIFIED TO SOME DEGREE.

即使没有这些金属离子，鱼类也不能在强酸性环境下生存……今天许多北方的湖泊已经在一定程度上呈酸性了。

OZONE DEPLETION

ATMOSPHERIC OXYGEN, EVERYONE'S FAVORITE GAS, NORMALLY EXISTS IN THE FORM OF A 2-ATOM MOLECULE, O_2.

臭氧消耗

大气中的氧气是大家最喜欢的气体，通常情况下它的存在形式是含有两个氧原子的氧气分子——O_2。

BUT HIGH ABOVE THE EARTH, EXOTIC CONDITIONS PRODUCE A 3-ATOM MONSTER CALLED ***OZONE***, OR O_3.

但是在远离地面的高空，受外太空条件的影响产生了一种特殊的三原子分子，叫作**臭氧**，也就是O_3。

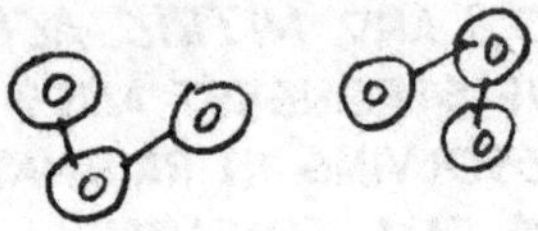

THIS LAYER OF OZONE PLAYS A CRITICAL ROLE IN MAINTAINING LIFE ON EARTH: IT SCREENS OUT THE SUN'S LETHAL ***ULTRAVIOLET (UV) RADIATION.***

臭氧层为保护地球的生命发挥着至关重要的作用：它能够屏蔽掉太阳发出的一些致命的**紫外线辐射**（UV）。

THEN ALONG CAME ARTIFICIAL ***CFCs***... SHORT FOR CHLOROFLUOROCARBONS. A CHEMIST'S DREAM, NON-TOXIC, NON-FLAMMABLE, HIGHLY STABLE COMPOUNDS, CFCs WERE FIRST SYNTHESIZED IN THE 1930s AND QUICKLY INCORPORATED INTO ***REFRIGERATOR COILS, AEROSOL SPRAY CANS***, AND THE BUBBLES IN ***STYROFOAM***. FROM THERE THEY FOUND THEIR WAY TO THE ***UPPER ATMOSPHERE***, WHERE THEY ***ATTACK OZONE.***

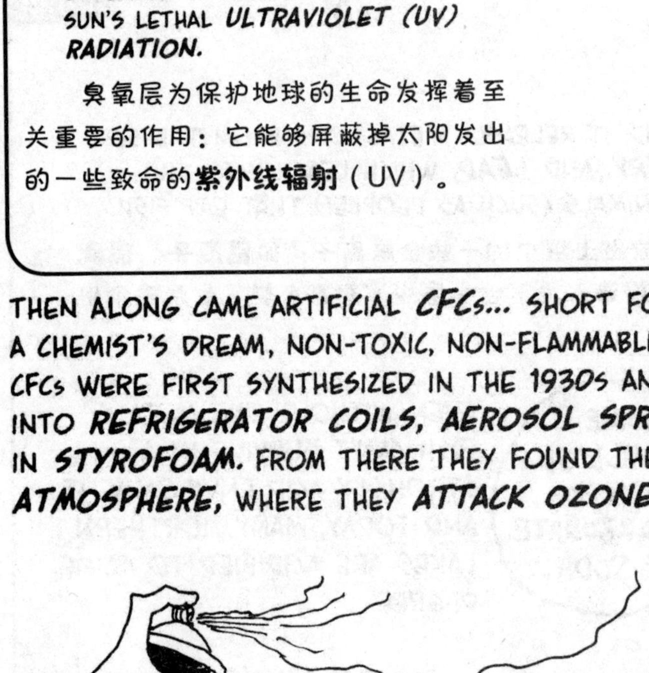

后来诞生了人造的CFCs,即氟氯烃。CFCs是化学家梦寐以求的一种无毒的、不燃的、稳定性很好的化合物，是在20世纪30年代首次被合成的，然后被迅速用于**冰箱线圈**、**喷雾罐**和**泡沫塑料**中。这种气体总能够进入**高层大气**中，**破坏臭氧**。

IN THE STRATOSPHERE, ***CHLORINE*** ATOMS ESCAPE FROM THE CFC MOLECULES... AND CHLORINE BREAKS DOWN OZONE MOLECULES, WHILE REMAINING UNAFFECTED ITSELF. ***A SINGLE CHLORINE ATOM CAN UNDO UP TO 100,000 MOLECULES OF OZONE,*** ALLOWING MORE UV RADIATION TO GET THROUGH TO THE EARTH.

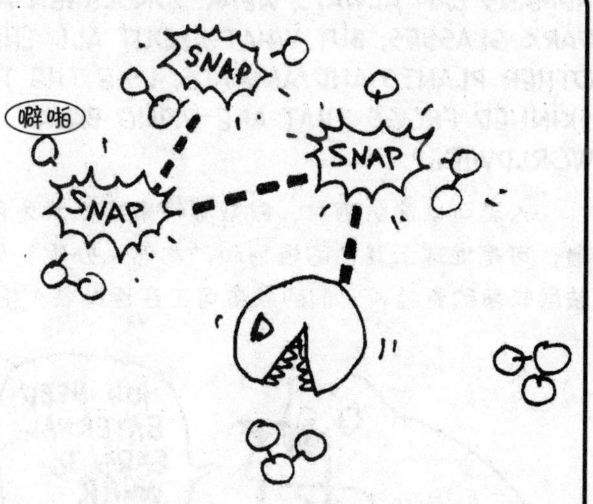

在平流层中，**氯原子**脱离氟氯烃分子，破坏臭氧分子，而氯原子本身却不受影响。**一个氯原子能破坏多达10万个臭氧分子**，从而使更多的紫外线到达地面。

SINCE THIS REACTION MAINLY HAPPENS ON COLD CLOUD SURFACES, THE OZONE "HOLE" FIRST APPEARED OVER THE SOUTH POLE... BUT OZONE IS ALSO THINNING OVER THE CONTINENTS... UV INTENSITY ON THE GROUND IS INCREASING, AND SO ARE ***SKIN CANCER RATES,*** EVEN AMONG DARKER-SKINNED PEOPLE, WHOSE PIGMENT HAD PREVIOUSLY PROTECTED THEM FROM MOST SOLAR RADIATION.

由于这种反应主要发生在气温比较低的云层表面，因此臭氧“空洞”首先出现在南极上空……但是地球陆地上空的臭氧层也在逐渐变薄……到达地面的紫外线强度逐渐增强，**皮肤癌的发病率**也增加了。甚至在一些黑皮肤的人群中皮肤癌的发病率也在逐渐增加，而他们皮肤中的色素本来可以帮助他们抵挡大部分来自太阳的紫外线辐射。

HUMANS CAN ALWAYS WEAR SUNSCREEN AND DARK GLASSES, BUT WHAT ABOUT ALL THE OTHER PLANTS AND ANIMALS, LIKE THE THIN SKINNED FROGS THAT ARE DYING BACK WORLDWIDE?

人类可以涂防晒霜、佩戴墨镜来抵挡紫外线辐射，可是地球上其他的植物和动物怎么办呢？如皮肤层较薄的青蛙，它们的数量可正在逐渐减少呢！

CFC PRODUCTION IS BEING PHASED OUT GLOBALLY, BUT REPLACEMENT CHEMICALS ARE NOT ALWAYS EASY TO FIND, AND THE STUFF ALREADY IN THE AIR CONTINUES TO EAT OZONE EVEN AS YOU READ THIS.

氟氯烃生产正在世界范围内逐渐被淘汰掉,可是也不容易找到用来替代它的化学物质,而且已经散发到空气中的氯氟烃要继续损耗臭氧。你在读这本书时，氯氟烃就正在损耗着臭氧。

AND FINALLY, THERE'S

GLOBAL WARMING

(PROBABLY!)

最后就是**全球气候变暖**的问题了。（很可能的！）

IF YOU TAKE ALL THE "POLLUTANTS" OUT OF EXHAUST GAS—THE SULFATES, NITRATES, LEAD, SOOT, ETC.—YOU ARE LEFT WITH PURE ***CARBON DIOXIDE,*** AND WHAT COULD BE WRONG WITH THAT? WE EXHALE THE STUFF ALL THE TIME, AND PLANTS BREATHE IT IN.

如果你把废气中的“污染物”（硫化物、硝酸物、铅、煤烟等）都处理掉，只剩下干净的**二氧化碳**，那二氧化碳能有什么问题呢？我们每天都会呼出这种气体，而植物会把它吸收掉。

AGAIN, THE PROBLEM IS SOLAR ***RADIATION.*** CARBON DIOXIDE GAS ABSORBS THE SUN'S RAYS AND WARMS UP... THAT BIT OF RADIATION IS NOT REFLECTED BACK INTO SPACE... AND SO, GENERALLY SPEAKING, THE MORE CARBON DIOXIDE IN THE AIR, THE HIGHER THE EARTH'S TEMPERATURE SHOULD BE.

又是因为太阳**辐射**的问题。二氧化碳能够吸收太阳光线因而使温度升高……而这部分被二氧化碳吸收的光线不会反射回太空中……因此，一般来讲，空气中的二氧化碳越多，地球的温度也就越高。

THE RESULT IS LIKE HEAT TRAPPED IN A GREENHOUSE, SO THIS IS CALLED THE ***GREENHOUSE EFFECT.***

结果就像一个能够收集热量的温室一样，因此这种效应被称作**温室效应**。

STUDIES OF ANCIENT EARTH CHEMISTRY SHOW THAT ATMOSPHERIC CO_2 LEVELS HAVE FLUCTUATED WIDELY OVER THE PLANET'S LONG HISTORY... AND THEY'VE BEEN STEADILY RISING FOR THE LAST CENTURY...

古地球化学研究显示，在漫长的地球历史长河中，空气中二氧化碳的水平波动幅度很大……但在20世纪，二氧化碳水平却呈不断增长的趋势……

SCIENTISTS ARE NOW TRYING TO SEE WHETHER ANCIENT RISES IN CO_2 CORRESPOND TO PERIODS OF WARMER CLIMATE.

科学家目前正在努力研究古时候二氧化碳水平增加的时期是否与气候变暖时期相吻合。

THERE IS MUCH DISAGREEMENT ABOUT GLOBAL WARMING... CLIMATE IS VERY IMPERFECTLY UNDERSTOOD... IT ISN'T EVEN CLEAR EXACTLY HOW TO MEASURE AVERAGE TEMPERATURES...

关于全球气候变暖有很多不一致的看法……气候本身就是一个很难全面理解的问题……甚至关于如何测量平均温度也不是很明确……

BUT SOMETHING SEEMS TO BE HAPPENING... SEA LEVEL HAS RISEN 6 INCHES SINCE 1900 (THE RESULT OF MELTING POLAR ICE? WATER EXPANDING FROM HEATING UP?)... HEAT WAVES ARE MORE COMMON... ALPINE PLANT ZONES ON MOUNTAIN SLOPES ARE STEADILY MOVING UPHILL, AS IF TEMPERATURES ARE BECOMING MORE TOLERABLE AT HIGHER ALTITUDES.

但似乎发生了一些异常的事情……自1900年以来，海平面已升高了6英寸*（是极地冰川融化导致的吗？温度升高导致了水的上涨？）……热浪变得更加普遍……半山腰上的高山植物带正逐渐向山上移动,似乎海拔越高的地区温度变得越适宜了。

*1英寸＝0.0254米，下同。

SCIENTISTS HAVE ALSO NOTICED THAT CARBON DIOXIDE IS NOT THE ONLY GREENHOUSE GAS. METHANE (CH_4), PRODUCED FROM MANY AGRICULTURAL SOURCES, FROM COW BURPS TO NATURAL FERMENTATION, MAY HAVE 20 TIMES THE EFFECT OF CO_2.

科学家也注意到，不只二氧化碳是温室气体。在农业生产过程中（从牛打嗝排气到自然发酵等）产生的甲烷（CH_4）,其温室效应可能是二氧化碳气体效应的20倍。

BY SOME ESTIMATES, WE CAN EXPECT A RISE OF ***2.5°*** CENTIGRADE BY THE YEAR ***2050*** (***4.5°*** FAHRENHEIT). THIS DOESN'T SOUND BAD UNTIL YOU REALIZE THAT IN THE LAST ***ICE AGE***, A ***4-DEGREE DROP*** PUT A ***MILE OF ICE*** UNDERFOOT. THINK ABOUT 20-FOOT HIGHER SEA LEVELS... OR 120° SUMMER DAYS...

据估计，到2050年，温度预期会上升2.5℃。这听起来或许并不太可怕，但如果你知道上个**冰期**，仅仅4℃**的降温**就在脚下产生了1英里的冰层，你就知道后果有多严重了。想一想，海平面若上升20英尺，或者夏天120°F(49℃)的高温……

ON THE OTHER HAND, IT MAY BE THAT HIGHER TEMPERATURES WILL CREATE MORE CLOUDS, WHICH REFLECT MORE HEAT... OR PERHAPS INCREASED PLANT GROWTH WILL ABSORB THE CO_2, KEEPING US COOL.

另一方面，高温或许会产生更多的云，这些云会反射更多的热量……或许加快生长的植物会吸收二氧化碳，使我们感到凉爽。

WHATEVER THE CASE, IT DOESN'T LOOK AS IF FOSSIL FUEL BURNING IS GOING TO SLACK OFF ANYTIME SOON... SO EXPECT CO_2 LEVELS TO KEEP RISING.

不管怎样，现在看起来化石燃料的燃烧不会很快减少……因此，二氧化碳的水平还会不断增加。

BY NOW, IT SHOULD BE CLEAR THAT WE INCREASINGLY LIVE IN A WORLD OF OUR OWN MAKING... AND THE QUESTION IS—WHAT ARE WE GOING TO DO ABOUT IT???

到此为止，无论你讨厌或是喜欢，我们目前的生活环境是由我们自己创造的……而问题是对此我们该怎么办呢?

CHAPTER 14

EARTH ISLAND

第十四章

地球岛

LIKE THE PEOPLE OF EASTER ISLAND, THE PEOPLE ON THE REST OF THE EARTH ARE RAPIDLY CHANGING THEIR ENVIRONMENT. ARE WE DOOMED TO A POPULATION CRASH AND A BARREN LANDSCAPE? OR CAN WE SUSTAIN A GREEN FUTURE FOR OURSELVES AND THE PLANET?

与复活节岛上的人们一样，地球上其他地方的人们也正在飞快地改变着他们的环境。我们会像复活节岛一样，终究摆脱不了人口崩溃、土地贫瘠的命运吗？还是可以为自己、为地球保留一个绿色的未来呢？

ONE OF THE SCARIEST FORECASTS OF DOOM
WAS GARRETT HARDIN'S 1968 ESSAY

"THE TRAGEDY OF THE COMMONS"

WHICH "PROVED" THAT DISASTER COULDN'T BE AVOIDED.

有关世界末日的最可怕的预言之一就是加勒特·哈丁在1968年发表的**《公地的悲剧》**一文。这篇文章"证明"了灾难是不可避免的。

IMAGINE A PASTURE, GOES HIS ARGUMENT, WHERE ***10 INDEPENDENT HERDERS*** GRAZE THEIR CATTLE. EACH HERDER WANTS TO MAXIMIZE HIS OWN WEALTH.

文章是这样论证的：假设有一片牧地，里面有10个**独立的牧民**在放牧。每个牧民都希望最大限度地增加自己的财富。

IF HERDER TOM ADDS *ONE COW* TO HIS HERD, WE CALL HIS GAIN *+1*.

如果牧民汤姆为他的牛群再添入一头牛，他的财富便增加了+1。

BUT THE ENVIRONMENTAL ***COST*** OF AN EXTRA COW IS SHARED BY ***ALL TEN HERDERS***, SO TOM'S COST IS ONLY 1/10. TO TOM, ***EVERY NEW COW IS PROFITABLE***.

但饲养这额外的1头牛需要消耗的环境成本由所有的10个牧民均分了，所以，汤姆在饲养这头牛上所花费的成本只有1/10。对汤姆来说，**每多养一头牛都是有利可图的**。

THE RESULT IS NO SURPRISE: TOM GETS AS MANY COWS AS HE CAN... AND SO DO DICK, HARRY, AND ALL THE OTHER HERDERS...
最后的结果无疑就是：汤姆尽可能多地养牛……然后什么迪克啊，哈利啊，其他的牧民也都这么干……
该死的，怎么突然就变得这么挤了？
DANG. HOW'D IT GIT SO CROWDED ALL OF A SUDDEN?

UNTIL THE COMMONS IS OVERGRAZED AND TURNS INTO A DESERT. IN THIS STORY, EVERYBODY LOSES.
直到这片公地因过度放牧而变成了荒漠。在这个故事里，每个人最终都是输家。
天啊。
OOPS.

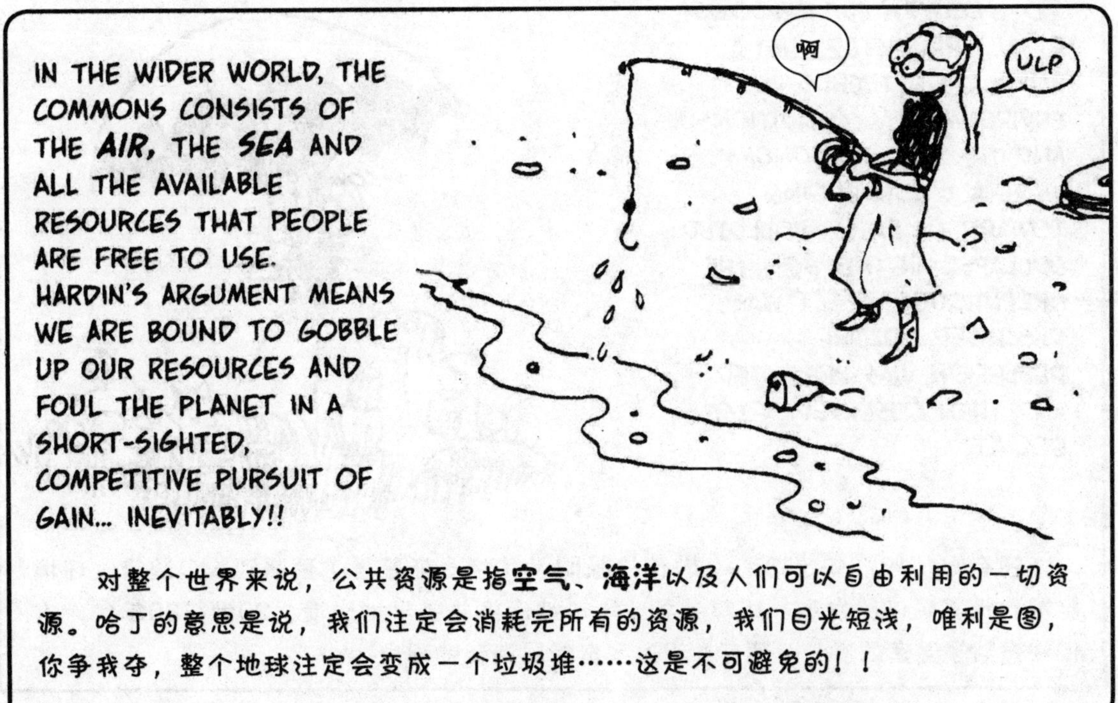
IN THE WIDER WORLD, THE COMMONS CONSISTS OF THE AIR, THE SEA AND ALL THE AVAILABLE RESOURCES THAT PEOPLE ARE FREE TO USE. HARDIN'S ARGUMENT MEANS WE ARE BOUND TO GOBBLE UP OUR RESOURCES AND FOUL THE PLANET IN A SHORT-SIGHTED, COMPETITIVE PURSUIT OF GAIN... INEVITABLY!!
啊
ULP
对整个世界来说，公共资源是指空气、海洋以及人们可以自由利用的一切资源。哈丁的意思是说，我们注定会消耗完所有的资源，我们目光短浅，唯利是图，你争我夺，整个地球注定会变成一个垃圾堆……这是不可避免的！！

ANOTHER CHEERFUL NOTE FROM 1968 WAS SOUNDED BY PAUL EHRLICH'S BOOK ***THE POPULATION BOMB.*** LIKE MALTHUS, EHRLICH SAW AN EXPLODING POPULATION AND CONCLUDED THAT MASS STARVATION WAS JUST AROUND THE CORNER.

另一个骇人听闻的预言来自1968年保罗·埃利希的著作**《人口大爆炸》**。像马尔萨斯一样，他看到人口在不断膨胀，因而推断大规模的饥荒可能就在眼前。

MORE? IN 1972, ***THE LIMITS TO GROWTH*** (BY MEADOWS ET AL.) PREDICTED THAT POPULATION TRENDS, ENVIRONMENTAL CONDITIONS, AND THE WORLD ECONOMY WERE ALL CONVERGING TOWARD AN EARLY, POLLUTED COLLAPSE. IN THE '80S, THE GREENHOUSE EFFECT WAS DESCRIBED... OZONE DEPLETION WAS PREDICTED AND THEN OBSERVED... ETC. ETC. ETC...

还有呢？1972年出版的**《增长的极限》**（梅多斯等著）预测到人口趋势、环境状况和世界经济的发展最终都会导致地球被污染并过早地崩溃。20世纪80年代，人们注意到了温室效应……预言并观察了臭氧层消耗的现象……

SO... WHAT, IF ANYTHING, IS WRONG WITH THESE ARGUMENTS?
那么……这些论点有任何不对的地方吗？
THERE IS SOMETHING WRONG, ISN'T THERE?
的确有不对的地方。
FIRST OF ALL, LET'S ADMIT IT: ALL THESE PESSIMISTS MAY BE RIGHT! WE HAVE THE EXAMPLE OF EASTER ISLAND, AND NATURE IS FULL OF POPULATIONS THAT HAVE COLLAPSED DOWN THE J-CURVE.
首先，我们得承认，所有这些悲观的看法都只是可能对的！比如复活节岛就是个例子，自然界也有很多种群呈J型曲线崩溃了。
YES... IT COULD BE HOPELESS...
是啊……可能毫无希望了……
AT THE SAME TIME, NO MATTER HOW TEMPTED WE ARE TO LIE DOWN AND WAIT FOR DOOMSDAY, THERE ARE SOME DISTURBING SIGNS OF... HOPE!
同时，无论我们怎么受悲观论点的影响，无论怎样无动于衷地躺下来被动等待世界末日的到来，但是还是有一丝……希望的迹象！
RATS... JUST WHEN I WAS GETTING COMFORTABLE...
瞎扯……我刚刚感觉舒服一点……

TAKE THE TRAGEDY OF THE COMMONS... (PLEASE!) THE FLAW THERE IS THAT HARDIN'S HERDERS ***NEVER TALK TO EACH OTHER...*** THERE IS NO ***MANAGEMENT*** OF THE COMMON LAND... THEY HAVE NO CONCEPT OF THE COMMON GOOD, NO VISION OF THE FUTURE.

拿公地的悲剧来说吧……(请好好想想！)问题在于哈丁书中的那些牧民们**彼此之间从不交流**……没有对公地进行**管理**……他们对共同利益没有概念，也对未来没有远见。

IN REALITY, MANY TRADITIONAL SOCIETIES HAVE MAINTAINED COMMON RESOURCES FOR HUNDREDS, EVEN THOUSANDS, OF YEARS. THE COMMONS MAY BE OVERSEEN BY A COUNCIL OF ELDERS, AND INDIVIDUAL COMPETITION IS RESTRAINED BY CUSTOM AND RELIGION.

现实生活中，许多传统的社会群体把公共资源维护了数百年甚至数千年。公共资源可能受到长老会的监管，并且人与人之间的竞争受到习俗和宗教的约束。

THE INDUSTRIALIZED WORLD HAS A MIXED RECORD. IN EASTERN EUROPE, WHERE COMMUNIST IDEOLOGY STRESSED INDUSTRIAL DEVELOPMENT AT ALL COSTS, THE ENVIRONMENT SUFFERED HORRIBLY. IN THE WEST, WHERE GOVERNMENTS PAY MORE HEED TO COMPETING INTERESTS, THERE ARE LARGE NATURAL PARKS AND ENVIRONMENTAL REGULATIONS—ALONGSIDE POLLUTION, OVERGRAZING, AND THE REST.

工业化世界在这方面喜忧参半。东欧国家强调不惜一切代价发展工业，因此环境遭受到了严重破坏。而在西方，政府更加注意相互的利益关系。虽然有污染、过度放牧等现象的发生，但也能看到许多大面积的自然公园，并有环保条例进行管理。

欢迎开车来国家公园

ON A WORLDWIDE SCALE, NO COUNTRY, COMPANY, OR ORGANIZATION OWNS THE OCEANS OR ATMOSPHERE, SO AT THIS POINT IT ISN'T EASY TO SAY HOW THEY CAN BE PROTECTED.

从世界范围来讲，海洋或大气层不属于任何一个国家、公司或组织，所以在这一点上，想说如何保护它们并不容易。

EVEN SO, THE NATIONS OF THE WORLD HAVE UNITED TO PHASE OUT OZONE-GOBBLING CFCs COMPLETELY AND TO REGULATE CERTAIN ASPECTS OF OCEAN EXPLOITATION. HOW THESE INTERNATIONAL AGREEMENTS CAN BE ENFORCED REMAINS TO BE SEEN.

即使这样，全世界的各个国家还是联合起来逐步淘汰了破坏臭氧层的氟氯烃气体的使用，并且对海洋开发的某些方面进行了管理。这些国际合约是否能有效实行还需要时间的检验。

AS FOR THE POPULATION BOMB—DESPITE POPULATION GROWTH, PER-CAPITA FOOD PRODUCTION HAS RISEN A LOT SINCE 1968. CHINA, INDIA, AND INDONESIA, WITH HALF THE WORLD'S POOR, BECAME SELF-SUFFICIENT IN GRAIN BY THE 1990s...

关于人口爆炸的问题，尽管人口持续增长，但自1968年起，人均粮食产量还是提高了很多。到20世纪90年代时，贫穷人口占世界一半的中国、印度以及印度尼西亚等国家也实现了粮食生产的自给自足……

AND MODERN BIOTECHNOLOGY STILL HAS A FEW TRICKS UP ITS SLEEVE.

现代生物技术还有许多锦囊妙计呢！

BUT WHAT ABOUT DEPLETING OUR RESOURCES? ALTHOUGH FARMLAND IS FINITE AND FOSSIL FUELS (EXCEPT COAL) WON'T LAST LONG, ALTERNATIVE FUELS DO EXIST, AND MOST NON-FUEL RESOURCES ARE STILL PLENTIFUL. MANY MINERALS CAN BE REPLACED BY CHEAPER, MORE ENVIRONMENTALLY FRIENDLY ALTERNATIVES.

可是资源消耗得这么快该怎么解决呢？虽然耕地面积有限，化石燃料（不包括煤）也不能维持很长时间，但替代燃料确实是存在的，而且大部分非燃料资源依然很充足。许多矿物都可以用成本更低、对环境更加友好的物质所代替。

THE "RESOURCE" WE ARE SQUANDERING MOST HEEDLESSLY IS ***THE BIOSPHERE ITSELF.*** WE DEPEND ON PLANTS TO TURN CO_2 INTO BIOMASS AND TO FILTER POLLUTANTS FROM THE AIR AND WATER. PLANTS REGULATE THE WATER CYCLE AND PLAY A CRITICAL ROLE IN OTHER CHEMICAL CYCLES AS WELL. OTHER ORGANISMS AERATE SOILS, STORE WATER, RECYCLE PLANT NUTRIENTS, CONTROL PESTS, POLLINATE FLOWERS, AND ON AND ON AND ON!!!

我们最不经意间消耗的“资源”就是**生物圈本身**。我们依赖植物将二氧化碳转化成生物量，并依靠植物来过滤空气和水中的污染物。植物主导着整个水循环过程，同时也在其他的化学循环中起着关键性作用。其他的生物给土壤通气，储存水分，循环利用植物的养分，控制害虫数量，为花朵授粉，如此等等，从不间断！！！

OUR RECKLESS USE OF THE BIOSPHERE IS THREATENING ITS ABILITY TO DO THE JOB WE REALLY NEED: RUNNING THE CHEMICAL CYCLES AND PRESERVING THE AIR, WATER, AND SOIL.

我们对生物圈毫无顾忌的利用正在削弱它的工作能力，使它不能满足我们真正的需求：使化学循环持续不断，保护空气、水和土壤。

SO WHAT IS A

SUSTAINABLE ECOSYSTEM?

ONE BIG LESSON OF THIS BOOK IS THAT ECOSYSTEMS ARE ***DYNAMIC.*** SUSTAINABLE DOES NOT MEAN DULL AND UNCHANGING. TO BE SUSTAINABLE, AN ECOSYSTEM NEEDS TO BE RESILIENT ENOUGH TO RECOVER FROM ACCIDENTS AND TO RECOLONIZE PREVIOUSLY LOST SPACE.

那什么是**可持续性生态系统**呢？我们最需要从这本书上学到的一点就是生态系统是**动态的**。可持续性并不意味着无精打采和一成不变。生态系统要想可持续，就需要有足够的恢复力，可以从意外事故中快速自我修复，并能够重新利用过去失去的空间。

SUSTAINABLE DEVELOPMENT SHOULD MEET THE NEEDS OF THE PRESENT WITHOUT DEPRIVING FUTURE GENERATIONS OF THE ABILITY TO MEET THEIR NEEDS. WE WANT TO LEAVE OUR CHILDREN A WORLD AT LEAST AS GOOD AS THE ONE WE WERE BORN INTO.

可持续性发展是应该在满足我们现有需要的同时，不剥夺后代满足他们需要的能力。我们想留给后代们的世界至少该像我们出生时的世界一样美好。

IN THE PAST, WE ALWAYS ASSUMED THAT OUR INGENUITY WOULD CREATE SOLUTIONS TO PROVIDE FOR FUTURE GENERATIONS, BUT CAN THIS GO ON INDEFINITELY? AT SOME POINT, WE HAVE TO FACE THE FACT THAT WE ARE JUST A PART OF A LIMITED BIOSPHERE. THERE IS ONLY SO MUCH MATTER AND ENERGY AVAILABLE!

在过去，我们常常认为凭借自己的智慧可以为子孙后代创造很多解决方案，但这是永久的吗？总有一天，我们必须面对一个事实，那就是：自己只是一个有限生物圈里的一分子。可用的物质和能量只有这么多！

IN THE LONG RUN, THIS CAN MEAN ONLY ONE THING: A ***NO-GROWTH ECONOMY,*** IN WHICH PEOPLE ENJOY A GOOD QUALITY OF LIFE BUT DON'T CONSUME SO MUCH STUFF.

从长远来看，这只有一个含义：那就是**无增长的经济**，即人们享受高质量的生活，但并不消耗这么多东西。

UNQUESTIONABLY, WE HAVE TO ***REDUCE WASTE:*** FARM WITHOUT EXCESS, IMPROVE EFFICIENCIES OF FUEL EXTRACTION AND ENERGY CONVERSION, AND REDUCE THE USE OF WOOD, METAL, AND PETROLEUM-BASED SYNTHETICS.

毫无疑问，我们必须**减少浪费**：适度耕种，提高燃料开采和能量转化效率，减少木材、金属以及石油合成品的使用。

IN AGRICULTURE, WE NEED TO ***INTENSIFY:*** GROW MORE CROP ON THE SAME LAND (OR EVEN LESS LAND, AS CITIES ENCROACH ON FARMLAND). AGAIN THIS IS A QUESTION OF LESS WASTE AND MORE CAREFUL HUSBANDRY OF PLANT NUTRIENTS.

在农业方面，我们需要精耕细作：在同样多的土地上（或者更少的土地，因为城市建设会侵占耕地）种植更多的农作物。这又是一个减少浪费而精心培植农作物、提高其营养成分的问题。

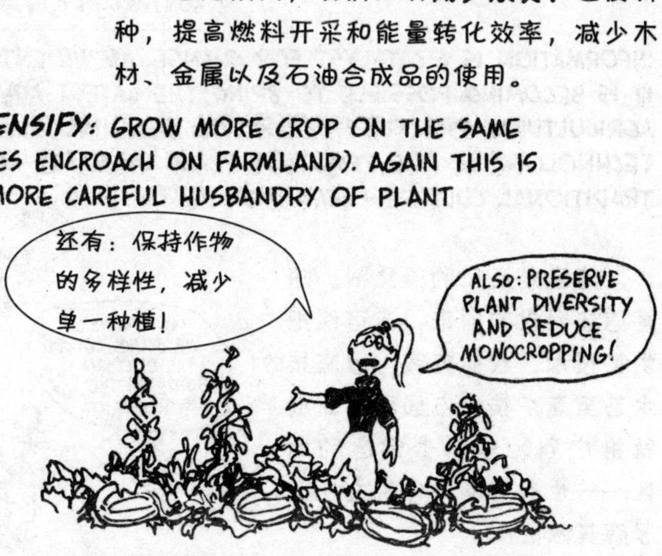

POPULATION CONTINUES TO GROW, AND NEARLY HALF THE NEW MOUTHS WILL BE IN CHINA, INDIA, PAKISTAN, BANGLADESH, AND NIGERIA, ALL COUNTRIES WITH WATER PROBLEMS AND DEGRADED ECOSYSTEMS. THEIR ECOLOGICAL RECOVERY IS UNLIKELY TO COME IN THE NEXT FEW DECADES.

人口持续增长着，而近一半的新增人口出现在中国、印度、巴基斯坦、孟加拉国和尼日利亚。这些国家都出现了水资源问题和生态系统退化问题。这些生态问题在日后的几十年里都未必能够恢复。

POPULATION CONTROL IS A GOVERNMENT GOAL IN MOST OF THOSE COUNTRIES. AS WE'VE SEEN, TWO OF THE MOST EFFECTIVE FORMS OF POPULATION CONTROL ARE PROSPERITY AND THE EDUCATION OF WOMEN.

在这些国家里，大多数都把控制人口作为政府工作的目标。正如我们看到的那样，两种最有效的人口控制方式就是让人民变得富裕和让妇女接受教育。

INFORMATION IS A CATALYST FOR CHANGE. AS WE ENTER THE INFORMATION AGE, IT IS BECOMING POSSIBLE TO BRING THE LATEST ADVANCES IN BIOTECHNOLOGY, AGRICULTURAL PRACTICES, IRRIGATION TECHNIQUES, AND EVEN COOKSTOVE TECHNOLOGY TO THE REMOTEST PLACE ON EARTH—AND TO LEARN FROM TRADITIONAL CULTURES HOW THEY DO IT!

信息是变化的催化剂。随着信息时代的到来，有可能把生物技术、农业实践、灌溉技术甚至烹炉技术方面的最新成就推广到地球上最偏远的地区——也有可能从传统文化中了解其做法！

FUNDAMENTALLY, GLOBAL CHANGE WILL COME FROM THE ***5+ BILLION INDIVIDUALS*** WHO LIVE ON THIS PLANET. USING LESS PAPER, METAL, AND PLASTIC... GROWING AND BUYING PESTICIDE-FREE FOOD... PLANTING TREES... CONSERVING WATER... USING LONGER-LIVED GOODS... RECYCLING AND BUYING RECYCLED PRODUCTS... HAVING FEWER CHILDREN... AND PUSHING OUR LEADERS TOWARD SUSTAINABLE POLICIES: ALL THESE ARE ELEMENTS OF GLOBAL CHANGE.

从根本上讲，全球变化取决于地球上居住的**五十多亿人口**。少使用纸、金属和塑料……种植和购买不喷洒农药的食物……植树造林……节约用水……使用耐用商品……回收利用并购买可回收利用的产品……少生小孩……推动领导人实行可持续发展政策：所有这些都是可以改变地球的做法。

WE ARE ALL PART OF THE BIOSPHERE ON THIS REMARKABLE ISLAND OF LIFE. EVERY INDIVIDUAL WHO "THINKS GLOBALLY AND ACTS LOCALLY" LIKE A RESPONSIBLE HETEROTROPH IS PART OF THE MOVEMENT TOWARD SUSTAINABILITY.

在这个奇妙的生命小岛上的生物圈里，我们都是其中的一部分。每一个有责任心的“从全球利益着想，从身边环境做起”的人，都应该做一个负责人的异养生物，为可持续发展做出贡献。

DON'T FORGET TO TURN OUT THE LIGHTS!

不要忘记关灯哦!

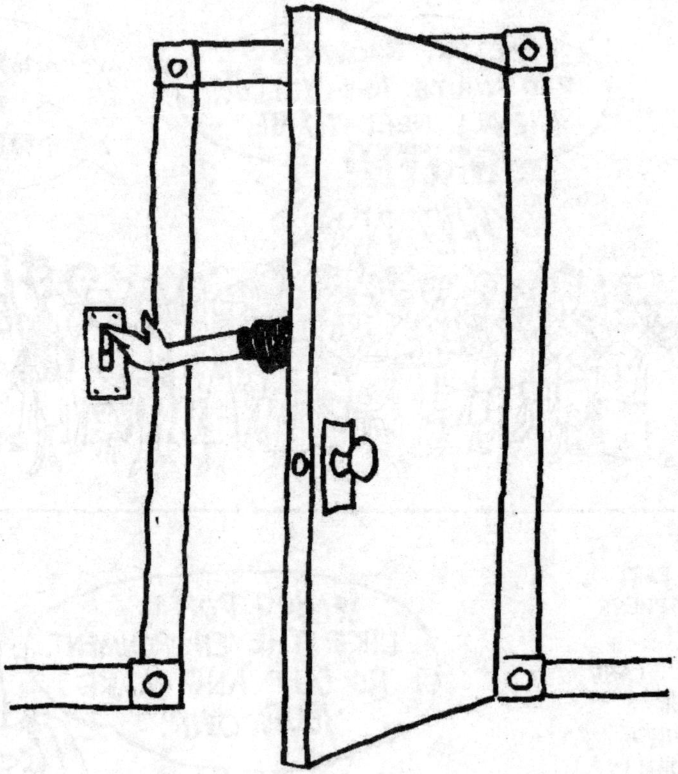

*本书采用环保纸印刷。

BIBLIOGRAPHY 参考文献

BAHN, PAUL AND JOHN FLENLEY. ***EASTER ISLAND, EARTH ISLAND.*** NEW YORK: THAMES AND HUDSON, 1992. 200 YEARS OF RESEARCH ON EASTER ISLAND, AND READABLE TOO.

CARTER, VERNON, AND TOM DALE. ***TOPSOIL AND CIVILIZATION.*** NORMAN, OKLAHOMA: UNIVERSITY OF OKLAHOMA PRESS, 1974. A CLASSIC ON DIRT.

CARTWRIGHT, FREDERICK. ***DISEASE AND HISTORY.*** NEW YORK: THOMAS Y. CROWELL, COMPANY, 1972.

DAY, DAVID. ***THE DOOMSDAY BOOK OF ANIMALS: AN ILLUSTRATED ACCOUNT OF THE FASCINATING CREATURES WHICH THE WORLD WILL NEVER SEE AGAIN.*** NEW YORK: VIKING PRESS, 1981. FROM BRITISH MUSEUM FILES, QUIRKY AND BEAUTIFULLY ILLUSTRATED.

DEBEIR, JEAN-CLAUDE, JEAN-PAUL DELEAGE AND DANIEL HEMERY. ***IN THE SERVITUDE OF POWER: ENERGY AND CIVILIZATION THROUGH THE AGES.*** LONDON: ZED BOOKS, 1991. UNBELIEVABLY ERUDITE, WITH A GREAT VIEW OF SLAVE SYSTEMS.

DEEVEY, E.S. "MAYAN URBANISM: IMPACT ON A TROPICAL KARST ENVIRONMENT." ***SCIENCE.*** OCT. 19, 1979: 298-306. MANY THEORIES COMPETE, BUT THOSE PHOSPHORUS SPIKES IN THE SEDIMENT CORES ARE MOST PERSUASIVE!

EHRLICH, PAUL R. AND ANNE H. EHRLICH. ***HEALING THE PLANET: STRATEGIES FOR RESOLVING THE ENVIRONMENTAL CRISIS.*** READING, MA.: ADDISON-WESLEY PUBLISHING COMPANY, 1991.

GIRARDET, HERBERT. ***THE GAIA ATLAS OF CITIES: NEW DIRECTIONS FOR SUSTAINABLE URBAN LIVING.*** NEW YORK: ANCHOR PRESS, DOUBLEDAY, 1992. WORLDWIDE VIEW OF CITY SYSTEMS.

HARDIN, GARRETT. "THE TRAGEDY OF THE COMMONS." ***SCIENCE.*** DEC. 13, 1968: (DON'T KNOW WHICH PAGES. SORRY!) SHORT, BUT IT MADE A BIG STIR.

HIESER, CHARLES B. ***SEED TO CIVILIZATION: THE STORY OF FOOD.*** CAMBRIDGE, MA.: HARVARD UNIVERSITY PRESS, 1990. GREAT STUFF IF YOU'VE NEVER SEEN IT, OTHERWISE ZZZZZZ.

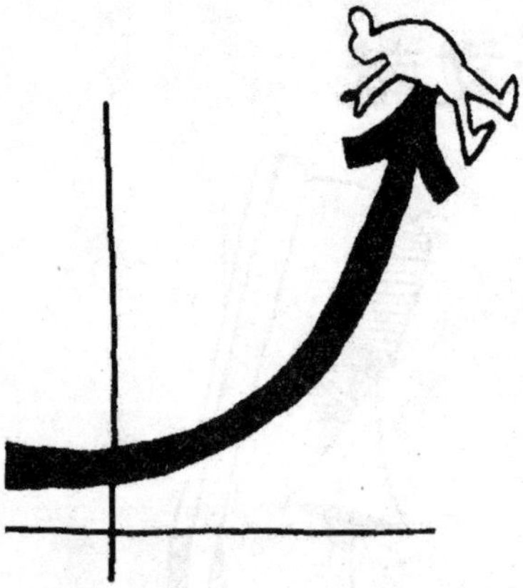

LOVELOCK, J.E. *GAIA: A NEW LOOK AT LIFE ON EARTH* NEW YORK: OXFORD UNIVERSITY PRESS, 1979. BRIEF AND PROVOCATIVE.

M^CNEILL, WILLIAM. *PLAGUES AND PEOPLES.* NEW YORK: ANCHOR BOOKS, 1976. NOT WELL WRITTEN, BUT THOUGHT-PROVOKING.

MEADOWS, D.H., D.L. MEADOWS AND J. RANDERS. *BEYOND THE LIMITS: CONFRONTING GLOBAL COLLAPSE, ENVISIONING A SUSTAINABLE FUTURE.* POST MILLS, VT.: CHELSEA GREEN PUBLISHING COMPANY, 1992. THE SEQUEL TO LIMITS TO GROWTH WITH LOTS OF DATA.

MEADOWS, D., ET AL. *LIMITS TO GROWTH.* NEW YORK: UNIVERSE BOOKS, 1972.

MILLER, G. TYLER. *LIVING IN THE ENVIRONMENT: SEVENTH EDITION.* BELMONT, CA.: WADSWORTH, 1992. FOR A COLLEGE TEXTBOOK ON ENVIRONMENTAL SCIENCE, IT CONVEYS A SURPRISING LEVEL OF FEELING ALONG WITH THE FACTS.

MYERS, NORMAN, ED. *GAIA: AN ATLAS OF PLANET MANAGEMENT.* NEW YORK: ANCHOR PRESS, DOUBLEDAY, 1984. A GOOD SYSTEMS OVERVIEW, BUT SOMETIMES OVERWHELMING GRAPHICS.

NEWMAN, LUCILE. *HUNGER IN HISTORY: FOOD SHORTAGE, POVERTY AND DEPRIVATION.* CAMBRIDGE: BASIL BLACKWELL, INC., 1990. LOTS OF SKELETAL STUDIES AND OLD FAMINE RECORDS.

PONTING, CLIVE. *A GREEN HISTORY OF THE WORLD: THE ENVIRONMENT AND THE COLLAPSE OF GREAT CIVILIZATIONS.* NEW YORK: ST. MARTIN'S PRESS, 1991. A BRILLIANT, DETAILED VIEW OF HUMAN IMPACTS ON EARTH.

RATHJE, WILLIAM AND CULLEN MURPHY. *RUBBISH! THE ARCHEOLOGY OF GARBAGE.* NEW YORK: HARPER PERENNIAL, 1992. LANDFILL DISSECTIONS AND GARBAGE REFLECTIONS.

WILSON, E.O. *THE DIVERSITY OF LIFE.* NEW YORK: W.W. NORTON AND COMPANY, 1992. EVERYTHING ABOUT BIODIVERSITY BY A MASTER.

ZINSSER, HANS. *RATS, LICE, AND HISTORY.* NEW YORK: BLUE RIBBON BOOKS, 1935. ECCENTRIC, ERUDITE, AND VERY FUNNY IN AN ODD, DRY WAY, THIS CLASSIC EXPLORES THE HISTORY OF TYPHUS.

ZISWILER, VINZENZ. *EXTINCT AND VANISHING ANIMALS: A BIOLOGY OF EXTINCTION AND SURVIVAL.* NEW YORK: SPRINGER-VERLAG, 1967. GREAT DOCUMENTATION OF THE FORMER TRADE IN ANIMAL PARTS.

ABOUT THE AUTHORS

作者简介

ALICE OUTWATER IS AN ENVIRONMENTAL ENGINEER TURNED ECOLOGY WRITER. A GRADUATE OF THE UNIVERSITY OF VERMONT WITH AN M.S. FROM MIT, SHE HELPED MANAGE THE SLUDGE FOR THE SIX-BILLION-DOLLAR BOSTON HARBOR CLEAN-UP. THIS IMMERSION IN THE OTHER END OF THE TOILET LED TO NUMEROUS TECHNICAL PAPERS AND A TEXTBOOK, ***REUSE OF SLUDGE AND MINOR WASTEWATER RESIDUALS.*** SHE LIVES WITH HER HUSBAND BOB AND THEIR SON SAM ON A FARM IN VERMONT, WHERE SHE WRITES AND CONSULTS IN WASTEWATER MANAGEMENT. HER THIRD BOOK, ***WATER: A NATURAL HISTORY,*** WILL BE PUBLISHED BY HARPERCOLLINS IN 1996.

嘿！萨姆，谷仓不能吃，家燕是名词。

爱丽丝·奥特瓦特是一名环境工程师，后来在生态领域著书立说。她本科毕业于佛蒙特大学，之后在麻省理工学院获理学硕士学位。她在波士顿港清洁项目的淤泥清除工作中发挥了重要作用，该项目经费高达60亿美元。爱丽丝·奥特瓦特对废水处理问题有深入的研究，发表了数篇技术论文，出版了题为《淤泥与废水残渣再利用》这一教科书。现在她与丈夫罗伯特和儿子萨姆居住在佛蒙特州的一个农场，在那里除了写作，她还从事废水处理的咨询工作。她的第三本书《水：自然历史》1996年由哈珀-柯林斯出版社出版。

ABOUT THE AUTHORS

作者简介

LARRY GONICK IS THE AUTHOR OR COAUTHOR OF NUMEROUS WORKS OF GRAPHIC NON-FICTION. TWO OF HIS BOOKS, ***THE CARTOON HISTORY OF THE UNIVERSE*** AND ***THE CARTOON GUIDE TO PHYSICS,*** HAVE BEEN ADAPTED TO CD-ROM. A CONTRIBUTING EDITOR OF ***DISCOVER*** MAGAZINE WITH THE BIMONTHLY FEATURE "SCIENCE CLASSICS," HE SPENT THE 1994-95 ACADEMIC YEAR AS A KNIGHT SCIENCE JOURNALISM FELLOW AT MIT. HE LIVES IN SAN FRANCISCO WITH HIS WIFE LISA AND HIS DAUGHTERS SOPHIE AND ANNA. FOR THEIR SAKE, HE HOPES YOU WILL ALL HELP TO SAVE THE ENVIRONMENT!

拉里·高尼克独立出版或与他人合著了多部图片著作，其中，《卡通宇宙历史》和《卡通物理指南》已制作成光盘。拉里·高尼克是《发现》杂志中“科学经典”栏目的编辑和撰稿人。1994~1995学年任麻省理工学院科学新闻学者，并在那里从事研究工作。现在他与妻子丽萨及两个女儿苏菲和安娜居住在旧金山。为了孩子们的利益，他希望我们大家都来拯救环境！

◆INDEX◆

索引

O

P

Q

R

S